Bioindicators of POPs

Bioindicators of POPs

Monitoring in Developing Countries

Shinsuke Tanabe

and

Annamalai Subramanian

Kyoto University Press

TRANS PACIFIC PRESS

First published in 2006 jointly by:

Kyoto University Press
Kyodai Kaikan
15-9 Yoshida Kawara-cho
Sakyo-ku, Kyoto 606-8305, Japan
Telephone: +81-75-761-6182
Fax: +81-75-761-6190
Email: sales@kyoto-up.gr.jp
Web: http://www.kyoto-up.gr.jp

Trans Pacific Press
PO Box 120, Rosanna, Melbourne
Victoria 3084, Australia
Telephone: +61 3 9459 3021
Fax: +61 3 9457 5923
Email: info@transpacificpress.com
Web: http://www.transpacificpress.com

Copyright © Kyoto University Press and Trans Pacific Press 2006

Printed in Melbourne by BPA Print Group

Distributors

QH
541.15
.I5
T36
2006

Japan
Kyoto University Press
Kyodai Kaikan
15-9 Yoshida Kawara-cho
Sakyo-ku, Kyoto 606-8305
Telephone: (075) 761-6182
Fax: (075) 761-6190
Email: sales@kyoto-up.gr.jp
Web: http://www.kyoto-up.gr.jp

Asia and the Pacific
Kinokuniya Company Ltd.
Head office:
38-1 Sakuragaoka 5-chome, Setagaya-ku,
Tokyo 156-8691, Japan
Phone: +81 (0)3 3439 0161
Fax: +81 (0)3 3439 0839
Email: bkimp@kinokuniya.co.jp
Web: www.kinokuniya.co.jp
Asia-Pacific office:
Kinokuniya Book Stores of Singapore Pte.,
Ltd.
391B Orchard Road #13-06/07/08
Ngee Ann City Tower B
Singapore 238874
Tel: +65 6276 5558
Fax: +65 6276 5570
Email: SSO@kinokuniya.co.jp

UK and Europe
Asian Studies Book Services
Franseweg 55B, 3921 DE Elst, Utrecht,
The Netherlands
Telephone: +31 318 470 030
Fax: +31 318 470 073
Email: info@asianstudiesbooks.com
Web: http://www.asianstudiesbooks.com

USA and Canada
International Specialized Book
Services (ISBS)
920 NE 58th Avenue, Suite 300
Portland, Oregon 97213-3786
USA
Telephone: (800) 944-6190
Fax: (503) 280-8832
Email: orders@isbs.com
Web: http://www.isbs.com

Australia and New Zealand
UNIREPS
University of New South Wales
Sydney, NSW 2052
Australia
Telephone: +61(0)2-9664-0999
Fax: +61(0)2-9664-5420
Email: info.press@unsw.edu.au
Web: http://www.unireps.com.au

ISBN 1-920901-11-6, 978-1-920901-11-0

b 2256576

Contents

Preface

Enhanced agricultural, vector control and industrial activities have resulted in a more comfortable and pleasant life for humans. Although such activities have helped bring about improvements in various facets of the quality of life, these involve increased usage of many man-made organic chemicals, resulting in many negative impacts, including environmental deterioration and health effects on wildlife. Most of these chemicals persist in the environment because of their resistance to almost all the known mechanisms of degradation of organic chemicals. They are transported over long distances, bioccumulated and biomagnified and cause a multitude of toxic effects. At present, the global environment is ubiquitously contaminated by a myriad of toxic chemicals, having teratogenic, carcinogenic, immunosuppressive and endocrine disruptive actions.

Twelve of these persistent organic pollutants (POPs) have been categorized as those that require immediate attention by a treaty adopted at a meeting of several nations held at Stockholm in 2001 and became legally binding in May 2004 after its ratification by 50 states. Although the treaty has been signed by 151 countries for its adoption, many of these countries are yet to ratify it. Moreover, exemptions have been provided to some states, particularly to the poor developing nations, for the controlled usage of some of these 12 chemicals for their disease-control practices until cost-effective alternative chemicals and/ or processes are found. At present, the developing nations have also become substantial sources of toxic compounds such as dioxins and related chemicals (DRCs) that are produced in relatively large amounts in their open dumping sites, leading to various environmental problems.

Such environmental problems are of great concern to both the developed and developing nations because almost all of these chemicals are semi-volatile and are distributed globally via various transport pathways. Pollution by these semi-volatile compounds always occur on a global scale and can never be regional. Therefore, all nations have the urgent duty of making an organized effort to evaluate the environmental changes caused by such chemicals and to control their negative effects. The development of easier, cost-effective and quicker methods to evaluate POPs in developing countries is long overdue. The easiest possible method is to use suitable bioindicators. The concentrations of POPs in several organisms have thus far been quantified in various parts of the world.

While the developed nations are making efforts to monitor and control the levels of pollution by all the 12 POP chemicals in their environment, the poor developing nations remain extremely hesitant to adopt planned monitoring; the prime reasons are the economy and the lack of man-power and facilities for such evaluations. The Global Environmental Facility (GEF) has recently included POPs into one of its operational programs. Therefore, all the countries that are parties to the convention will certainly need a book such as this for guidance in implementing the Stockholm Convention.

All those who are involved in the monitoring of POPs—the policy makers, scientists, research students, laboratory technicians, university students and non-governmental institutions—need such a book at present. The material included in this book is concise yet precise.

We have designed the book primarily as a broad treatment of this discipline which would be accessible to almost all the people working on POPs. Many currently available books contain a chapter(s) on the usage of bioindicators for measuring different pollutants. However, to our knowledge, this will be the first book that is exclusively dedicated to 'Bioindicators of Pollution by POPs'. This is a timely publication that is presently required by scores of people belonging to scientific, academic and governmental organizations as well as NGOs engaged in studies on POPs. Many laboratories in developed nations such as USA, Japan and the EU have now shown interest in evaluating the contaminant levels of POPs in their own countries as well as in the developing nations around them. Hence, this book will certainly serve as a guideline for scientists from both the developed and developing countries. We are proud to present this book as our small contribution to the scientific community involved in the research on POPs, and we believe that this book may help, at least in a small way, to relieve the suffering faced by the general public and wild animals due to the toxic manifestations of POPs and several other persistent toxic substances.

Acknowledgements

This book is the essence of the efforts of several people involved in research on POPs worldwide. We sincerely thank the entire scientific fraternity and other scholars affiliated to various governmental and non-governmental institutions who have contributed information regarding this field of study. This book would not have been possible but for the efforts of individuals who have contributed a significant amount of information on this subject.

We are extremely grateful to the Scientific and Technical Advisory Panel (STAP) of the Global Environmental Facility (GEF) and the United Nations Environment Program (UNEP) for their valuable assistance by various means during the preparation of the initial manuscript. We thank Dr. John Giesy, Department of Zoology, Michigan State University, USA, and Dr. Kurunthachalam Kannan, New York State Department of Health and Department of Environmental Health and Toxicology, State University of New York at Albany, USA, for critically reviewing the entire manuscript, despite their tight schedules and also for their valuable suggestions and ideas for improving the quality of the book. We also thank Dr. Derek Muir, National Water Research Institute, Environment Canada, Canada, and Dr. Paul K.S. Lam, Department of Biology and Chemistry, City University of Hong Kong, Hong Kong SAR, China, for their invaluable suggestions and ideas during the different stages of preparation of the manuscript.

We are thankful and have great respect for Prof. Masayuki Komatsu, President of Ehime University, and Prof. Hidetaka Takeoka, Director, Center for Marine Environmental Studies (CMES), Ehime University, for providing encouragement and valuable guidance. Our thanks are also due to Emeritus Prof. Ryo Tatsukawa, our guide and mentor, for all the knowledge that we have gained from him in this field.

The publication of this book was financially supported by Grants-in-Aid for Publication of Scientific Research Results (No. 175302) from the Japan Society for Promotion of Science and the 21st Century Center of Excellence Program (COE) of the Ministry of Education, Culture, Sports, Science and Technology, Japan. We gratefully acknowledge their financial assistance.

We thank the teaching staff, research scholars and students of the Division of Environmental Chemistry and Ecotoxicology, CMES, Ehime University, for all their assistance during the preparation of this book. We also thank the administrative staff of CMES and Ehime University.

But for the efforts of the staff of Kyoto University Press and Trans Pacific Press, this book would not have been delivered on time. Our special thanks are due to all their staff members.

Shinsuke Tanabe
Annamalai Subramanian
Center for Marine Environmental Studies (CMES),
Ehime University
Bunkyo-cho 2-5,
Matsuyama 790-8577,
Japan

About the Authors

Dr. Shinsuke Tanabe is acclaimed as one of the world's best environmental chemists and ecotoxicologists in the field of persistent toxic substances (PTS). He has authored nearly 300 original scientific publications and 60 book chapters and articles, both in English and Japanese. Presently, he is a Professor at the prestigious Center for Marine Environmental Studies (CMES), Ehime University, Japan. He is the Programme Leader of the 21st Century Center of Excellence Program sponsored by the Ministry of Education, Culture, Sports, Science and Technology of Japan as well as the Asia Pacific Mussel Watch Project. He has worked in collaboration with international organizations such as the STAP/GEF and has also worked as a consultant on persistent organic pollutants (POPs) for several other institutions. The exquisite Environmental Specimen Bank at CMES, catering to the needs of many research institutions, is his brain child. He has received innumerable research awards in Japan and abroad, including the Okada Prize from the Oceanographical Society of Japan (1985), the Nissan Science Prize from the Nissan Science Foundation (1999), the Citation Classic Award in Japan from ISI Thomson Scientific (2000), the Friendship Award for Collaborative Academic Activities from the Government of Vietnam (2003), the Academic Achievement Awards from the Japan Society for Environmental Chemistry (2004), the Society of Environmental Science, Japan (2004) and the 2005 SETAC (Society of Environmental Toxicology and Chemistry) Founders Award.

Dr. Annamalai Subramanian was born in India and has nearly 30 years of experience in pollution studies on persistent organic pollutants (POPs). After working as a Teaching and Research Faculty at Annamalai University, India, for nearly 20 years, he is presently working as a Guest Professor of Marine Environmental Chemistry at the Center for Marine Environmental Studies (CMES), Ehime University, Japan. He has obtained fellowship awards from the Japanese Ministry of Education, Culture, Sports, Science and Technology and the Japan Society for Promotion of Science. He has authored more than 100 published scientific articles on the subject. Apart from handling several research projects on POPs in India, he has conducted scientific studies for the United Nations University, Tokyo, Japan, the Toyota Foundation, Japan and the STAP/ GEF. He is now a Roster Expert of UNEP for assessing projects on POPs.

Abbreviations Used in This Book

AhR	Aryl hydrocarbon receptor
AMAP	Antarctic Monitoring and Assessment Program
APMW	Asia Pacific Mussel Watch Program
BCF	Bioconcentration factor
CB	Chlorobiphenyl
CDV	Canine distemper virus
CEC	Criteria Expert Group
CHL	Chlordane
CHLs	Chlordane compounds
CISEM	Centro per l'innovazione e la sperimentazione educativa Milanese (Workshop Series on Mediterranean Mussel Watch Project)
CLS	Commission on Life Science
CR	Concentration ratio
CYP	Cytochrome P450
DDD	1,1-dichloro-2,2-bis(4-chlorophenyl)ethane
DDE	1,1-dichloro-2,2-bis(4-dichlorodiphenyl)ethylene
DDT	1,1,1-trichloro-2,2-bis(4-dichlorodiphenyl)ethane
o, p'-DDT	1,1,1-trichloro-2-(2-chlorophenyl)-2-(4-chlorophenlyl) ethane
p,p'-DDT	1,1,1-trichloro-2,2-bis(4-chlorophenyl) ethane
DDTs	DDT and its metabolites
ECOD	7-ethoxycoumarin-O-deethylase
EMAP	Environmental Monitoring and Assessment Program
EOM	Extractable organic matter
EPA	Environmental Protection Agency
EROD	Ethoxyresorufin-O-deethylase
FAO	Food and Agriculture Organization
GEF	Global Environment Facility
GNP	Gross National Product
HCB	Hexachlorobenzene
HCH	Hexachlorocyclohexane
HCHs	Hexachlorocyclohexane isomers
INC	Intergovernmental Negotiating Committee
IFCS	Intergovernmental Forum for Chemical Safety
IUPAC	International Union for Pure and Applied Chemistry

MC	Methylchlonthrene
MeSO$_2$-PCBs	Methylsulfonyl PCBs
MeSO$_2$-DDE	Methylsulfonyl DDE
MI	Metabolic index
NADP	Nicotine adenine dinucleotide phosphate
NADPH	(Reduced form of) Nicotine adenine dinucleotide phosphate
NCBP	National Contaminant Biomonitoring Program
NOAA	National Oceanic and Atmospheric Administration
NPMP	National Pesticide Monitoring Program
NRC	National Research Council
NS&T	National Status and Trends
OCs	Organochlorines
PAHs	Polyaromatic hydrocarbons
PB	Phenobarbital
PCBs	Polychlorinated biphenyls
PCDDs	Polychlorinated dibenzo-*p*-dioxins
PCDFs	Polychlorinated dibenzofurans
PCN	Polychlorinated naphthalene
PCP	Pentachlorophenol
POPs	Persistent organic pollutants
PROD	Pentoxyresorufin-*O*-deethylase
PTS	Persistent toxic substances
STAP	Scientific and Technical Advisory Panel
TCPMe	*tris*(4-chlorophenyl)methane
TCPMOH	*tris*(4-chlorophenyl)methanol
TEQs	Toxic equivalents
U-DGA	Urinary D-glutaric acid
UNDP	United Nations Development Program
UNEP	United Nations Environment Program
UNESCO	United Nations Education, Scientific and Cultural Organization
UK	United Kingdom
US	United States
USA	United States of America
USSR	Union of Soviet Socialist Republics
WHO	World Health Organization

Entry of POPs into the Environment

Under the Stockholm Convention, 12 chemicals were designated as persistent organic pollutants (POPs) for their immediate control. The main routes of entry of these chemicals into the environment are via their use in pesticides, industries and electrical installations. In addition, uncontrolled burning of municipal wastes in open dumping sites is becoming a major source of dioxins and related chemicals.

Mussels: Universal Pollution Indicators

Bivalve mollusks have been extensively used and proved successful as bioindicators for monitoring POPs in natural waters. These organisms, particularly mussels and oysters, possess all the characteristics required of a bioindicator for POPs. Brackish water species belonging to the genera *Perna*, *Mytilus* and *Crassostrea* were found to be most suitable. Freshwater mussels can be used for monitoring inland areas.

Squids: Indicators of Pollution by POPs in Open Oceans

Squids are carnivores and have simple food habits. There are 375 squid species in the oceans around the world; none of these migrate over long distances. Squids transfer contaminants through gills by equilibrium partitioning. The concentrations of toxicants in their bodies would reflect the pollution levels in the seawater at the location and time of their collection; thus, they are good bioindicators for measuring ambient levels of pollution by POPs.

Fishes Reflect Pollution by POPs in all Water Bodies

Fishes are ubiquitous and occur in almost all water bodies. Most fish species are good accumulators of persistent and lipophilic compounds such as POPs and reflect the environmental levels of these compounds in most instances. Fishes can be used as bioindicators of POP exposure in international waters after careful consideration of their various life history parameters.

Birds: "Spies" of Local, Regional and Global Pollution

Birds are useful as bioindicators of POPs in certain cases. While the resident birds may reflect the background pollution in their habitat, migratory birds can act as 'spies' of the pollution levels in their wintering and feeding grounds as well as along their migratory routes. With careful sampling strategies, the global pattern of pollution by POPs can be integrated using birds as bioindicators.

Marine Mammals: Indicate Global Pollution by POPs and their Toxic Effects

The recent mass strandings and epizootics in marine mammals are correlated with high concentrations of POPs such as DDTs and PCBs in their bodies. Seas and oceans are final reservoirs of POPs. Being at the top of the marine food chain, marine mammals accumulate heavy loads of these contaminants in their lipid-rich tissues. Thus, they become suitable bioindicators of pollution by POPs.

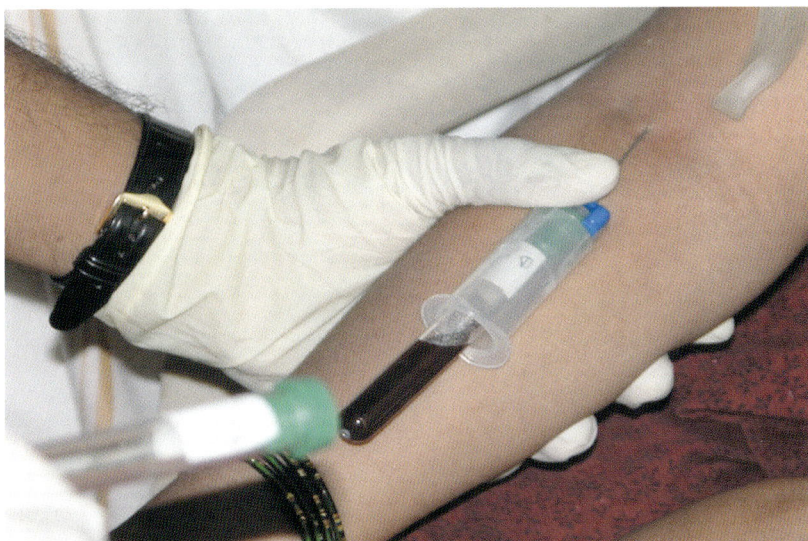

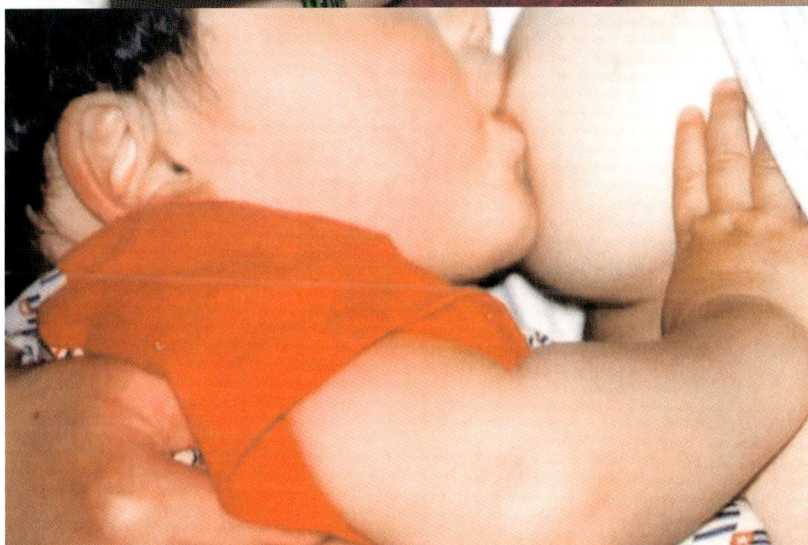

Humans: Bioindicators Providing Most Relevant Data on Pollution

Humans are exposed to POPs through a variety of routes—air, water and food. Despite some social, ethical and legal impediments, human tissue samples collected in a non-destructive manner are the best samples for measuring spatial and temporal variations in POPs. The best advantage of using human tissue samples is that background data such as age, reproduction and possible routes of exposure can be accurately obtained from the subject(s).

Mussels: Preferred as Bioindicators of POPs over Other Animals
Mussels are always preferred as bioindicators for many pollutants because of their ability to accumulate a myriad of contaminants. They can withstand varying environmental conditions and bioaccumulate contaminants to higher levels through filter-feeding. They also possess all the characteristics required of a good bioindicator of POPs.

Entry of POPs into the Environment

Under the Stockholm Convention, 12 chemicals were designated as persistent organic pollutants (POPs) for their immediate control. The main routes of entry of these chemicals into the environment are via their use in pesticides, industries and electrical installations. In addition, uncontrolled burning of municipal wastes in open dumping sites is becoming a major source of dioxins and related chemicals.

Chapter 1: Introduction

Profile

Overview

Bioindicators or animal sentinels are the individuals of a population or a group of animals that are used indirectly to measure the levels of contaminants in their ambient environment; bioindicators are systematically collected and analyzed to identify potential health hazards to other animals and humans. These sentinel organisms can be grouped according to the parameter they are designed to monitor (e.g. exposure or effect), the types of animals used, the environment in question or whether the animals are in their natural habitat (observational systems) or are deliberately deployed into an environment in question (experimental or *in situ* systems) (CLS, 1991).

For monitoring persistent organic pollutants (POPs), the systems may be designed to reveal environmental contamination, food chain transfer of contaminants or to investigate the bioavailability of contaminants from environmental media (bioindicators). Alternatively, the systems may be designed to provide an early warning of contamination by a particular contaminant or a group of contaminants (biomarkers). Some sentinel organisms can be used to indicate both exposure (biomonitors) and hazards (biomarkers). While the term 'bioindicator' has been in existence since the 19th century, the term 'biomarker' has been gaining acceptance in recent years, although there are some inconsistencies in the definitions of both these terms.

In this book, we have used the term 'bioindicators' to describe the long-term 'exposure' of an animal at an 'organismic' level, that is, the accumulation of POPs in individual tissues and organs of the animal(s). Effects and responses at 'sub-organismic' levels of the organisms, termed as 'biomarkers' does not come under the preview of this book and will be discussed only at appropriate places, whenever needed.

Descriptions by various authors have made the definitions and divisions of bioindicators and biomarkers very ambiguous. We opine that the old definitions of bioindicators provided by Goldberg (1975) and Phillips (1980) remain valid for our purpose of recommending suitable bioindicators to developing countries for measuring the POP contamination levels in terrestrial, coastal and oceanic ecosytems.

By definition, bioindicator studies comprise chemistry-based analyzes of the concentrations of bioaccumulated contaminants in the tissues and organs of indicator organism(s) and the behavioral changes in individuals, populations and communities. In this review, we have restricted ourselves to evaluating the available literature on the POP concentrations recorded in various animal species worldwide, discussing their merits and demerits and recommending guidelines for selecting a universal bioindicator to monitor POPs in coastal and oceanic ecosystems, the ultimate reservoirs of pollutants used in the terrestrial environment.

Purpose

As one of the integral operational programs of the Global Environment Facility (GEF) that works with and through the United Nations Environment Program (UNEP), the global treaty and ratification for the ban on 12 POPs received special emphasis on the agenda. As part of this agenda, the GEF has requested its Scientific and Technical Advisory Panel (STAP) to review and evaluate the globally available data on the 12 POPs (that are designated for immediate action in the Stockholm Convention signed in 2001). Most of these POPs are either banned or under severe restriction in many of the developed nations, but some developing countries continue to use them in agriculture and for disease control. Because of the prohibitive costs of POP analysis, which also requires the use of sophisticated instrumentation for measurements, many developing countries are reluctant to undertake studies on the contamination caused by these chemicals. These countries are becoming potential sources of global pollution by POPs due to their indiscriminate use of these chemicals, inadequate management of abandoned manufacturing facilities, stockpiles, improper burning of municipal wastes at open dumping sites, etc. Therefore, it is imperative to undertake steps to discover simple, cost-effective and accurate methodologies as well as suitable bioindicator species and biomarker response studies that can be effectively used by the developing countries for routine POP monitoring, particularly in their existing laboratories. In the three intended aims of the STAP/GEF Workshop on the Use of Bioindicators, Biomarkers and Analytical Methods for the Analysis of POPs in Developing Countries, held at Tsukuba, Japan this book will review the available knowledge regarding bioindicator studies on POPs involving the use of different aquatic organisms and recommend suitable bioindicator methodologies for the future to meet the needs of developing countries.

Structure

This book has been written after contemplating on the available information regarding the concentrations and distribution of POPs in different animal species in order to recommend appropriate bioindicator organism(s) for monitoring of these chemicals in the developing countries in the future.

This chapter is an overview on the production and usage of second-generation pesticides and other organochlorines (OCs) as well as their effects on various non-target organisms. In addition to definitions and explanation of the bioindicator concept, this chapter describes the need for the use of bioindicators in regular monitoring surveys. It also illustrates the essential characteristics required by an organism to be a bioindicator for POP monitoring. The subsequent chapters contain compiled information on several animals that have been used to monitor POPs in several nations.

Chapter two illustrates the use of mussels as sentinels for POPs, cites various studies that have been conducted so far on the distribution of POPs in different mussel species in relation to ambient media and describes their temporal and spatial considerations. The chapter also discusses the suitability of these animals as bioindicators of POPs, the problems that may be encountered during monitoring surveys and the ways to overcome such problems.

The third chapter describes the use of squids as sentinels for POP monitoring. Available literatures pertaining to the use of each squid species have been compiled and examined to evaluate the merits and demerits as well as to recommend appropriate approaches for using squids for POP monitoring. The fourth, fifth and sixth chapters describe studies on the potential applicability of fish, birds and marine mammals, respectively for POP monitoring.

Chapter seven reviews studies on the occurrence of POPs in human tissues, organs and body fluids. In addition to discussing humans as bioindicators, this chapter also emphasizes the necessity and importance of survey on effects of POPs on human subjects. The eighth and final chapter illustrates the use of animal sentinels in risk assessment with a focus on the criteria for selection and application of suitable animal sentinel(s) for spatial and temporal monitoring. This chapter provides recommendations and discusses some guidelines for efficient application of bioindicators.

Each chapter explains the potential use of different groups of animals as bioindicators (chapters two to seven). These chapters start with a brief introduction; this is followed by a description of factors that may affect the bioaccumulation of POPs, previous attempts on using those animals as bioindicators of POPs and concluding remarks. Cited literatures are listed at the end of each chapter for easy reference.

Outline

Environmental Contamination

Prior to the Second World War, plant protection was dependent on the use of a few inorganic substances, namely, common compounds of arsenic, copper or sulphur, complemented by a few naturally occurring organic chemicals such as nicotine and pyrethroids as insecticides. Many organic compounds were developed as second-generation pesticides for military and public use during and after the Second World War. After the war, there was a rapid development of the chemical industry. The resulting development of OCs opened a new era in pest control. The use of chemicals for pest control has increased since the discovery of the pesticidal properties of 1,1-dichloro-2,2-bis(p-dichlorodiphenyl)ethane (DDT). It was very effective in controlling pests on agricultural crops and malaria-spreading mosquitos. This was followed by the invention and widespread use of similar chemicals such as hexachlorocyclohexane (HCH), aldrin, dieldrin, heptachlor and other chemicals, many of them having chlorine as one of their constituents.

In addition to pesticides, a myriad of other industrial chemicals were produced and released either intentionally or accidentally. Organic chemicals such as polychlorinated dibenzo-p-dioxins (PCDDs) and polychlorinated dibenzofurans (PCDFs) detected in environmental and biotic media are mostly produced unintentionally. These chemicals are emitted as impurities into the environment due to incomplete combustion of hospital and municipal wastes and during bleaching of pulp by using chlorine in paper industries as well as during the manufacture of pesticides and other chlorinated substances such as pentachlorophenol (PCP) and polychlorinated biphenyls (PCBs).

Most of these chemicals, which were used in the last few decades of the 20th century, possessed characteristics that include a broad spectrum of activity, persistence and insolubility in water. These qualities that favored their use as pesticides rendered them most undesirable from the environmental viewpoint. Later, the use of persistent organic chemicals for agriculture, disease control and industries became ubiquitous and thousands of persistent chemicals have become firmly established in several aspects of our life.

Hundreds of POPs exist at present; these often belong to certain 'families' of chemicals. For example, theoretically, there are 209 PCB congeners (which differ from each other in their levels of chlorination, ranging from monochlorinated PCBs that contain only 1 chlorine atom in a biphenyl ring to decachlorinated isomers that contain 10 chlorine atoms), 75 isomers and congeners of dioxins and 135 dibenzofurans—all differentiated by chlorine numbers and positions.

POPs are highly persistent, having a long half-life in the environment and biota. They may have a half-life of years or decades in soils or sediments and several days in the atmosphere (Jones and Voogt, 1999). POPs are typically hydrophobic and lipophilic. In the aquatic environment, they partition strongly to the organic phase. The POPs dissolve in lipids and accumulate in the fatty tissues of animals, rather than entering into the aqueous milieu of cells. Hence, these chemicals persist in the biota and biomagnify in food chains.

Another important characteristic of POPs is their semi-volatility; therefore, they enter into the atmosphere under normal ambient temperatures. In the atmosphere, they adhere to aerosols, rather than residing in the gaseous phase. Hence, they may volatilize from the earth into the atmosphere and travel long distances in the air before being redeposited. This cycle of volatilization and deposition, which has been termed as 'grasshopper effect' (Wania and Mackay, 1993), may be repeated several times, resulting in the transfer of POPs to an area distant from their source.

At present, these chemicals are ubiquitous; they are carried away by air and surface runoff to distant regions, including the remotest parts of the globe. Depending on their transportability, the existing climatic conditions and atmospheric circulation patterns, these chemicals have been distributed across various regions of the globe (Tanabe et al., 1994). The transport pathways include air, ocean currents, runoff into surface waters, leaching through soils into sub-surface waters and animal migrations (Rolland et al., 1995).

During the last two decades, the scientific community has become increasingly aware of the short-term and long-term effects of POPs on the environment, plants, animals and humans. In its report in 1999, the World Wildlife Fund has stated that only approximately 70–90% of the pesticides applied on the ground and 25–50% of the sprayed pesticides reach their target areas, and the remainder are transported into various compartments of the environment; this exposes all forms of life, including humans and wildlife, to these pesticides (WWF, 1999).

POPs are ubiquitous; they are found in the most remote areas, distant from any agricultural, industrial or disease control activities. The occurrence of some of these chemicals in Antarctica has been reported in studies conducted as early as the 1960s (Sladen et al., 1966). Although they occur in the environment at very low concentrations, POPs have been linked to many health and environmental effects. Surface waters are the final reservoirs of all such chemicals and substances that are naturally drained from terrestrial ecosystems or produced by human activities.

The ecological condition of estuaries and coastal waters is affected by a complex array of stressors. These may include point sources such as pollutant loading from a variety of direct point sources; general discharges from surface and ground waters; non-point runoff from agricultural, urban and other systems and from atmospheric deposition. Summers et al. (1997) stated that by the end of the 20th century, 75% of the US population might be living within 50 miles of the coastal area, thereby increasing the already significant threat to estuarine and coastal ecosystems. At present, increase in the percentage of coastal population is a worldwide phenomenon, both in the developed and developing countries. Hence, these areas represent a suitable environment to assess the utility of biomonitoring techniques for evaluating the biological impacts of natural and anthropogenic stresses on aquatic organisms.

POPs are continued to be used in some of the developing countries. The existing climatic conditions characterized by high atmospheric temperature and torrential rainfall highly favor the volatilization and transport of POPs via the atmosphere and/or transport to nearby seas and inland waters via surface runoff or through leaching from the soils (Ramesh et al., 1990; Takeoka et al., 1991; Iwata et al., 1993, 1994). Depending on their transportability and volatility characteristics, POPs travel long distances. They remain suspended in the atmosphere and are transported thousands of kilometres away from their source, travelling across national boundaries. The portion that reaches the seas and oceans is further distributed by surface and subsurface currents.

Apart from this, vast stockpiles of POPs have been maintained in many developing countries over the last four decades. Some of these stockpiles have been exported by developed countries in the past (Loganathan and Kannan, 1994). These stockpiles of chemicals may pose a threat to the global environment in the future. These chemicals may leach out or evaporate slowly from these stockpiles and may be globally distributed via various transport pathways.

The last decades of the 20th century have witnessed a steady increase in the extent and pace of environmental change. Concerns over issues such as biodiversity loss, atmospheric pollution, changes in water quality and land usage pattern and climatic change as well as their impacts have highlighted the need for high quality, reliable long-term data to interpret environmental trends and to assist in policy-making (Parr et al., 2003).

The presence of DDT in human tissues was known as early as the mid-1940s. Among the first reports on OC residues in wildlife were those regarding the presence of DDT in avian tissues (Mohr et al., 1951), followed by the recovery of aldrin residues from birds (Post, 1952), soil (Gannon and Bigger, 1958) and on plants (Gannon and Decker, 1959). The scope and intensity of pesticide contamination were defined largely based on residue data that documented the fate of insecticides in the environment. It became clear that residues were not distributed equally in all habitats or animals. These facts led to the conclusion of existence of some environmental 'hot spots'—areas that either received repeated treatments or continuous input. Studies supported by residue analyzes showed that POPs were transported from treated areas via air and water. This dispersal of residues was responsible for the contamination of animals even in remote and isolated areas such as the Antarctic (Subramanian et al., 1983; Tanabe et al., 1986; Aono et al., 1997), Arctic (Addison and Smith, 1974; Muir et al., 1995, 2000; Nakata et al., 1998; AMAP, 1998) and almost the entire global ecosystem (Risebrough et al., 1968; Tanabe, 2002; UNEP, 2003).

Several reviews have been published on POP residues in terrestrial and aquatic animals and their ecosystems (Moore, 1965; Stickel, 1973). These reviews considered the types and amounts of contaminants in animal tissues and their possible significance. A book by Edwards (1973) titled 'Environmental Pollution by Pesticides' critically evaluated the residue levels in the environment and their effects on animals. Such studies helped biologists to understand the causes of pesticide-related mortalities and potential threats of pesticides to wildlife populations. The worldwide pesticide contamination threatened the wildlife and affected their productivity and hence their biodiversity (Stickel, 1973).

All such concerns related to POPs resulted in an initiative that culminated in the signing of the POPs treaty after a series of five consecutive meetings in Johannesburg, South Africa, in 2000. The POPs treaty will, for the first time in history, eliminate or severely restrict the use and production of a pernicious group of chemicals that are directly toxic to wildlife, ecosystems and people. This treaty as well as the Stockholm Convention on Persistent Organic Pollutants, or the POPs Convention, held in Stockholm in 2001 aims to protect human health and the environment from 12 chemicals. These chemicals are of particular concern due to their four intrinsic characteristics, namely, wide spectrum (thus affecting non-target organisms), persistence (thus resulting in bioaccumulation), transportability (thus causing global pollution) and toxicity (thus causing adverse effects). The 12 POP chemicals, known as the 'dirty dozen', covered by the POPs Convention include pesticides such as aldrin, chlordanes (CHLs), mirex, DDT, toxaphene, dieldrin, endrin and heptachlor; industrial chemicals such as PCBs and unintentionally produced chemicals such as hexachlorobenzene (HCB), which has application as a fungicide and synthetic intermediate), PCDDs and PCDFs. Based on sound scientific information, each of these chemicals has been linked to adverse human and wildlife health effects, including cancer, nervous disorders, damage to reproductive system and disruption of endocrine system. Many of these chemicals are also known to cause deleterious environmental effects.

Although the GEF has already adopted some measures to restrict or curtail the global release of POPs, a considerable amount of work remains to be accomplished. The Stockholm Convention also aims at banning and destroying the world's most dangerous chemicals and managing the stockpiles of wastes containing POPs in a safe manner, taking into account international guidelines and, in particular, the needs of the developing countries (http: //www. africastockpiles.org/treaties.html).

The GEF also aims at developing environmentally sound projects for the monitoring and safe disposal of such stockpiles. Ironically, most of these stockpiles are in the world's poorest regions that cannot afford the cost of even monitoring, let alone safe disposal. Apart from African countries, several thousand tons of such chemicals, mostly in unsafe conditions, were reported to be found in various types of intentionally and unintentionally accumulated stockpiles in some countries in Asia, Europe, South and North Americas and the former Soviet countries.

Apart from stockpiles, open dumping sites in many of the developing countries are posing a different type of environmental threat. While most of the developed nations have highly efficient incinerators for the disposal of their domestic and industrial wastes, many developing countries continue to dump their everyday wastes in open dumping sites on the outskirts of major cities; the dumping is often unregulated. The intentional (by waste pickers) and unintentional burning (through generation of methane gas) of wastes produces toxic chemicals such as dioxins and related chemicals; these chemicals may be transported by various pathways and cause global pollution.

Thus far, no step has been taken by these countries to control the open dumping of wastes mainly due to economic reasons. Because of the number and extent of dumping sites, these poor nations may pose a real threat of global pollution by toxic chemicals. Recent surveys in some Asian developing countries such as Cambodia, India, Vietnam and Philippines suggested that dumpsites are potential sources of pollution by dioxins, furans and PCBs (Minh et al., 2003).

These important reasons form the basis of the GEF's operational programs such as the International Waters Program, and for POPs being included in one of the special operational programs on its agenda working with and through the UNEP, one of GEF's implementing agencies. POPs will be receiving special emphasis in the overall GEF agenda. The STAP of the GEF has decided to gather information on POPs by declaring them as one of the prime areas of concern. The STAP conducted workshops to prepare explanatory manuals for discussion and documents on POPs for consideration by the GEF. Bioindicators, biomarkers and analytical methods suitable for developing countries were the three areas of prime importance discussed in one of the workshops.

Stockholm Convention
Of all the pollutants released into the environment through human activity, in either industry or agriculture, and disease control, POPs are the most dangerous. Over the last few decades, these highly toxic chemicals have killed and injured wildlife by inducing cancer and damaging the nervous, reproductive and immune systems. They have also caused several birth defects.

Traces of these chemicals may be found in all humans. POPs are highly stable compounds that can persist for years or decades before breaking down. They circulate globally through a process known as the 'grasshopper effect'.

Through a repeated process of evaporation and deposition, POPs released in one part of the world can be transported through the atmosphere to regions that are distant from their source.

In addition, POPs are concentrated in living organisms through another process called bioaccumulation. Although they are not soluble in water, POPs are readily absorbed in fatty tissues where their concentrations may magnify by up to 70,000 times the background levels. Fish, predatory birds, mammals and humans are high up in the food chain; therefore, they absorb the greatest POP concentrations. Further, the POPs are transported to different regions due to migration of these organisms.

Because of these two processes, even humans and animals living thousands of kilometres away from any major POP source, for example, in the Arctic and Antarctic regions, accumulate high POP concentrations in their bodies. However, POPs are equally dangerous to people working with pesticides or living near POP sources, particularly in developing countries, where a lack of equipment and expertise leads to accidental exposures.

Although some of these chemicals were banned in many developed and some of the developing countries since the 1970s, the fact that these chemicals can 'jump' via the atmosphere and can also be transported across borders through other environmental and biotic media necessitated a globally binding agreement, applicable and acceptable to all nations. Such an agreement would have to be legally binding on all countries and at the same time consider peculiar individual socioeconomic and political constraints, which may differ between countries. Under the present circumstances, there is no alternative to a legally binding international document for controlling the use of POPs.

In May 1995, the UNEP Governing Council requested in its Decision 18/32 that an international assessment might be undertaken on an initial list of 12 POPs. The chemicals selected for immediate action are aldrin, CHL, DDT, dieldrin, dioxins, endrin, furans, HCB, heptachlor, mirex, PCBs and toxaphene. The UNEP instructed the Intergovernmental Forum on Chemical Safety (IFCS) to develop recommendations on international action, no later than 1997, for consideration by the UNEP Governing Council and World Health Assembly. In June 1996, the IFCS provided sufficient recommendations, including a globally binding legal document, to call for international action to reduce risks to human health and environment due to the release of the 12 POPs.

In February 1997, the UNEP Governing Council invited the UNEP to convene an Intergovernmental Negotiating Committee (INC) to prepare a legally binding instrument for implementing international action, initially focussed on the 12 POPs. It also requested the INC to establish an expert group to develop criteria and a procedure for identifying additional POPs as candidates for future international action. The first meeting of the INC was held in June 1998 in Montreal, Canada; the Criteria Expert Group (CEG) was established during this meeting. Subsequent meetings of the INC were held in Nairobi, Kenya, in January 1999; Geneva, Switzerland, September 1999; Bonn, Germany, March 2000 and Johannesburg, South Africa, December 2000, when negotiations were successfully completed. Further, the CEG drafted its mandate in two meetings; the first was held in Bangkok, Thailand, in October 1998 and the second, in Vienna, Austria, in June 1999. In June 2000, the INC convened a meeting of 18 countries in Vevey, Switzerland, for discussing the financial resources required for the effort.

The convention was adopted and opened for signature at a conference held on 22 and 23 May 2001 in Stockholm, Sweden. On 23 May 2001, it was initially signed by 92 states. It remained open for signature from 24 May 2001 to 22 May 2003 at the UN Headquarters in New York; during this period, the number of signatory countries increased to 151. It was announced that the convention would enter into force 90 days after the submission of the 50th instrument of ratification by respective governments.

Initially, 30 articles and 6 annexures were ratified and signed by the participating nations, and the UNEP and INC were empowered to oversee the implementation of the international action on the convention and to take interim actions, until the convention came into force. The 90-day countdown to the treaty's entry into force was triggered on 17 February 2004 when France became the 50th state to ratify the agreement. At Geneva/Nairobi on 18 February 2004, the UNEP announced that the convention would become legally binding on 17 May 2004. As of 23 July, 2005 the convention has been ratified by 103 countries.

Governments pursued a rapid start to action under the treaty when they met for the first session of the Conference of the Parties to the Convention (COP 1) in Punta del Este, Uruguay, during 2-6 May 2005. The meeting succeeded in adopting a broad range of decisions, including guidance on national implementation plans (NIPs) and technical assistance, and the establishment of the POPs review committee (POPRC). One of the priorities of this meeting was the concerns expressed by the developing countries about their capacity to comply without technical assistance, in particular regarding best available technologies (BAT) and best environmental practices (BEP) to reduce dioxin and furan, two unintentionally produced POPs.

The main intention of the convention is to immediately ban most of the 12 chemicals. However, DDT use may be exempted for disease vector control under WHO guidelines because in many countries, DDT use remains essential for controlling malaria transmission by mosquitos. This will permit governments to protect their citizens from malaria—a major killer in many tropical regions—

until DDT can be replaced with chemical and non-chemical alternatives that are cost-effective and environmentally friendly. Meanwhile, the convention will help direct research and development towards more effective means of malaria control.

In addition to banning the use of such chemicals, the treaty also focuses on cleaning up the growing numbers of unwanted and obsolete stockpiles of pesticides and toxic chemicals. Consideration should be given to the elimination of obsolete stockpiles of pesticides in some developing countries. By committing governments to the elimination of production and environmental releases of these chemicals, the Stockholm Convention will greatly benefit human health and the environment. It will also strengthen the overall scope and effectiveness of international environmental law.

Yet another key goal of the convention will be to finalize guidelines for promoting 'best environmental practices' and 'best available techniques' for the reduction and elimination of releases of dioxins and furans (perhaps the most toxic of all the POPs) from a wide range of industrial and other sources such as open dumpsites for municipal and other wastes in developing countries. In the case of PCBs, governments should phase out the uses of PCB-containing electrical equipments by introducing PCB-free replacements. The governments must dispose of these PCBs in an environmentally friendly manner no later than 2028.

Fortunately, alternatives to POPs are available. The high costs, lack of public awareness and absence of appropriate infrastructure and technology have often prevented their adoption. It was generally accepted in the convention that suitable solutions must be tailored to the specific properties and uses of each chemical as well as to climatic and socioeconomic needs of each country.

To ensure that such solutions are exploited, donor countries have pledged to contribute hundreds of millions of dollars in new funding over the next several years. The GEF is the principal entity of the interim financial mechanism of the treaty. It has already mobilized resources to support POP-related projects such as the African Stockpiles Program in more than 100 countries. Backed by an alliance of developed and developing countries and having both industry and environmental groups on board, the Stockholm Convention holds the promise of a POPs-free world for future generations.

Dirty Dozen- Target Chemicals of Stockholm Convention

By definition, POPs are organic compounds that are highly resistant to degradation by chemical, biological and photolytic means. These substances are characterized by low water solubility and high lipid solubility, remarkable environmental persistence, long half-lives and bioaccumulative potential; they travel long distances via the atmosphere. Twelve POPs, popularly known as the 'dirty dozen', have been slated for immediate phasing out and removal under the Stockholm Convention. A brief summary of each of these chemicals has been provided for better understanding before delving into the details of their biomonitoring studies.

Aldrin

Aldrin is normally used to control termites, worms, weevil and grasshoppers as well as for protecting crops such as corn and potatoes against various soil insects. Aldrin is readily metabolized to dieldrin by plants and animals. Aldrin readily binds to sediments, and it rarely leaches into ground water. It may be volatilized from the sediments and redistributed by air currents, contaminating areas distant from their sources. Acute aldrin poisoning can kill birds and fish as well as humans. Chronic aldrin poisoning occurs through exposure to contaminated food. This compound has been banned in most of the developed nations, but continues to be used as a termiticide in many developing countries in Asia and Africa. Although highly toxic, aldrin is also very species specific. It can cause liver damage, which was reported as its primary effect (Orris et al., 2000), but it was also reported to be carcinogenic and cause neurological and reproductive disorders. It is banned in many countries, including Bulgaria, Ecuador, Hungary, Singapore, Switzerland, etc. and severely restricted in many countries, including Argentina, Canada, Japan, New Zealand, USA, etc.

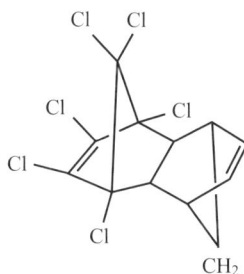

Chlordane (CHL)

Chlordane is a broad-spectrum contact insecticide used on agricultural crops and for the control of termites. It is highly insoluble in water but is soluble in organic solvents and is semi-volatile. It binds readily to aquatic sediments and bioconcentrates in the fatty tissues of organisms. Human exposure occurs mainly through air. Its acute toxic effects are species specific. CHL is thought to be a cancer and tumour promoter and may cause slight neurological disorders. It can mimic sex steroids or change their levels in exposed individuals. CHL usage has been banned in several countries, including Brazil, Netherlands, Philippines, Singapore, Spain, Sweden, etc. Its use is severely restricted or limited to non-agricultural use in Argentina, Canada, China, New Zealand, USA, etc.

DDT
DDT was perhaps the most infamous of the POPs; it was initially used during the Second World War to protect troops and civilians from malaria, typhus and vector-borne diseases. After the war, DDT was widely used for agriculture and disease control. It has a strong persistence in soil. Being the earliest, well known and one of the most widely used pesticides, DDT caused widespread contamination of water and soil resources, resulting in serious health effects in humans and animals. Because of its effectiveness as an insecticide, the WHO has developed an action plan to balance an ultimate phase out while endorsing a limited government-authorized use in public health campaigns and indoor residual applications (World Health Assembly, Resolution No. 50. Eighth plenary meeting, 12 May 1997). DDT is readily metabolized into a stable and equally toxic compound 1,1-dichloro-2,2-bis(p-dichlorodiphenyl)ethylene (DDE). Food is the primary route of exposure to DDT and its metabolites. Both these compounds are fat soluble and accumulate well in adipose tissues. In animals, DDT causes adverse health effects such as reproductive and developmental failure and may cause immune system defects. As with many other OCs, the nervous system is the major target of acute DDT exposure. Chronic exposure leads to neurological, hepatic, renal and immunological effects. As of 1995, DDT has been banned in 34 countries and severely restricted in an additional 34 countries (Ritter et al., 1995). After its ban, DDT residues have declined steadily over the past 20 years; however, DDT residues are continued to be detected in all environmental and biotic samples, including human breast milk, which raises potential concerns regarding effects on infants.

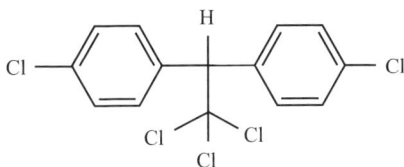

Dieldrin

Dieldrin has been principally used in agriculture for the control of soil insects and several insect vectors of disease. The pesticide aldrin rapidly converts to dieldrin in the environment and in animal bodies; therefore, dieldrin concentrations in the environment are higher than those indicated by dieldrin use alone, and the dieldrin residues that are found in tissues are likely to be an additive effect of aldrin and direct dieldrin exposure. Dieldrin residues have been found in air, water, soil, fish, birds and mammals, including humans. Food represents the primary route of exposure in the general population. Dieldrin use has been banned in many countries, including the EU, Singapore, Sweden, etc. and has been severely restricted in Argentina, Canada, Austria, Columbia, India, New Zealand, USA, etc.

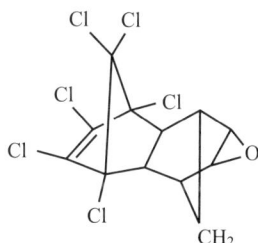

Endrin

Endrin is a rodenticide that is used to control mice and voles. It is sprayed on the leaves of crops such as cotton and grains. Unlike other OC compounds with similar structures, endrin is rapidly metabolized by animals, and it does not accumulate in fatty tissues to the extent as other related compounds; however, endrin is more acutely toxic than the closely related aldrin and dieldrin. It is highly toxic to fish. Food is the primary route of exposure in the general public. Hepatic and fetal abnormalities as well as liver and brain damage have been reported in endrin-treated rats, mice and hamsters. Endrin is banned in many countries such as Finland, Israel, Philippines and Singapore and severely restricted in Argentina, Canada, Chile, India, Japan, Pakistan, USA, etc.

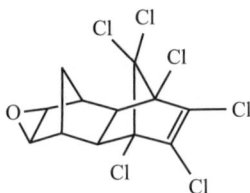

Heptachlor
Heptachlor has been used to kill cotton insects, grasshoppers and malaria-carrying mosquitoes. It is a stomach and contact insecticide. It is very volatile and highly insoluble in water but is soluble in organic solvents. It binds strongly to aquatic sediments, bioconcentrates in fatty tissues and partitions into the atmosphere. Heptachlor epoxide, the metabolic product of heptachlor in animal bodies, is equally toxic and is as bioaccumulative as the parent compound. It is highly toxic to humans and causes hyperexcitation of the central nervous system and liver damage. Effects of heptachlor on progesterone and estrogen levels were also observed in rats. It has been classified as a possible human carcinogen. Food is the major route of exposure in humans. Use of heptachlor has been banned in more than a dozen countries, and its use is severely restricted in countries such as the EU, Canada, Japan, New Zealand, USA, etc.

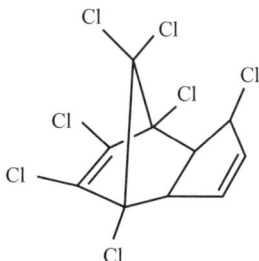

Hexachlorobenzene (HCB)
Hexachlorobenzene (HCB) is a fungicide that was used for seed treatment in the late 1940s. It is a by-product of the manufacturing process of industrial chemicals such as carbon tetrachloride, perchloroethylene, trichloroethylene and pentachlorobenzene; and it is an impurity in many pesticide formulations. HCB has been found in all types of food. In high doses, HCB was found to be lethal to some animals, and in lower doses, it caused skin lesions in human and severe reproductive damage in many animals. HCB is very persistent and is transported over long distances. It has a high partition coefficient, leading to its high bioconcentration.

HCB has been banned in Austria, Belgium, Denmark, the EU, Netherlands, Panama, UK, etc. and is restricted in Argentina, New Zealand, Sweden, etc.

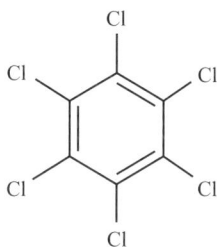

Mirex

Mirex is a bait insecticide with little contact activity; it is mainly used against ants, wasps, bugs and termites. It has industrial applications; it is used as a fire retardant in plastics, rubber, paint and electrical goods. In the presence of light, mirex breaks down to photomirex, a far more potent toxin. It is one of the most stable and persistent pesticides. It is very resistant to breakdown, is insoluble in water and has been found to bioaccumulate and biomagnify. Mirex binds strongly to aquatic sediments. There is limited data on human injuries caused by exposure to mirex. It is an endocrine disrupter, immunosuppressor, carcinogen and teratogen in animals. Humans are exposed to mirex mainly through food, particularly meat, fish and wild game.

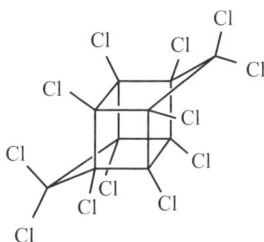

Polychlorinated biphenyls (PCBs)

Polychlorinated biphenyls (PCBs) are mixtures of chlorinated hydrocarbons. Since the 1930s, PCBs have been widely used in industrial applications, for example, in electrical transformers and capacitors, paint additives, copy papers and plastics. Theoretically, 209 PCB isomers are possible; these may range from three monochlorinated isomers to the fully chlorinated decachlorobiphenyl isomer, with decreasing water solubility and vapour pressure and increasing lipid solubility with increasing chlorine substitution. In practice, commercial mixtures available under various brand names in different countries contain varying numbers of isomers. PCBs are normally released into the environment as an impure mixture containing other chemicals. They have low volatility and are relatively persistent, chemically stable and resistant to heat; this creates an environmental problem. PCBs without an ortho substitution are generally referred to as coplanar PCBs; the others, non-coplanar PCBs. Coplanar PCBs, which are 13 in number, exhibit a dioxin-like toxicity. PCBs are insoluble in water; they tend to adhere to organic particles in the environment and get bioaccumulated in the fatty tissues of animals. Food is the main route of exposure in humans. The US Environmental Protection Agency (EPA) classifies PCBs as a human carcinogen. They are neurotoxic and affect foetuses. Non-human species, including rats, mice, monkeys and quail, have also exhibited clear neural changes resulting from PCB exposure. Apart from these, the hormone disrupting effects of PCBs are profoundly evident.

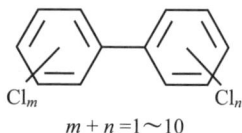

$$Cl_m \qquad Cl_n$$
$$m + n = 1 \sim 10$$

Polychlorinated dibenzo-*p*-dioxins and furans (PCDDs and PCDFs)
Polychlorinated dibenzo-para-dioxins (dioxins) and polychlorinated diben-zofurans (furans) are two groups of planar tricyclic compounds that have very similar chemical structures and properties. Neither dioxins nor furans are produced commercially, and they have no known commercial use. These chemicals are produced unintentionally due to incomplete combustion of hospital and municipal wastes as well as during the manufacture of several chlorinated compounds. There are 75 different dioxins, and furans have 135 positional isomers. At least 20 of these compounds are considered potentially toxic. Approximately 90% of the human exposure occurs through food, particularly animals. Daily exposure results in their accumulation in lipids, breast milk and blood. Because mothers' milk is often highly contaminated, infants are exposed to higher doses. The overall toxicity of a dioxin-containing mixture is assumed to be the toxic equivalent (TEQ) of a stated amount of pure 2,3,7,8,-tetrachlorodibenzo-*p*-dioxin (TCDD)—the most potent, hazardous and well studied dioxin—(Cohen et al., 1997). At present, only chloracne has been shown as the persistent effect associated with these chemicals in humans. Further, through the dysregulation of genes, dioxins can directly affect the growth and differentiation of cancer-producing cells. Dioxins are well known endocrine disrupters and immunosuppressors.

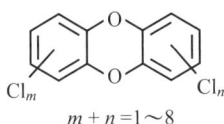

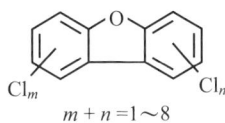

Cl_m Cl_n Cl_m Cl_n
$m + n = 1 \sim 8$ $m + n = 1 \sim 8$

Toxaphene

Toxaphene is a non-systemic and contact insecticide used on cotton, fruits, etc. and for controlling mites and ticks that infest farm and domestic animals. Toxaphene is essentially non-toxic to plants. As in the case of other pesticides, the low water solubility, high stability and semi-volatility of toxaphene favor its long-range transport. Although toxaphene usage has been stopped since 20 years, some countries permit its use in small quantities. It induces altered enzyme activities in the liver and has dose-dependent effects on the kidneys, thyroid and nervous system. It has been reported to display some estrogenic effects and to be associated with cancer in mammals. This chemical has been banned in several countries, including Egypt, the EU, India, Korea, Singapore, etc., and its use has been restricted in Argentina, Pakistan, South Africa, Turkey, etc.

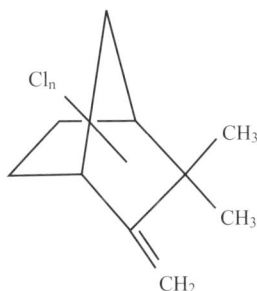

After identifying the above-mentioned 12 POPs, the Stockholm Convention took a precautionary approach for acting against other chemicals with characteristics of POPs. This provision calls for the evaluation of chemicals based on the criteria of toxicity, persistence, bioaccumulation and long-range transport for inclusion of these chemicals in the treaty. As stated in a section of the treaty, there is always a need for precaution; lack of scientific certainty should not prevent a POP from being included in the list. Therefore, it is always imperative to study the chemicals with the same characteristics as POPs. Following this, some authors have already recommended some chemicals such as polybrominated diphenyl ethers (PBDEs) (Ueno et al., 2004) for inclusion in the list of POPs.

Acknowledging the need for such precautions, we have cited examples from the works of several authors in which they have also studied chemicals other than the 12 POPs. While discussing certain chemicals along with POPs, some authors preferred to use the term persistent toxic substances (PTS) (Barakat, 2004). Although some of the OCs having all the characteristics of POPs are not included in the list of POPs, these chemicals have been evaluated by many authors (e.g. HCH). Throughout this book, while citing such chemicals, we have used the same terminology used by the authors of those publications, for example, POPs, PTS and OCs. All these chemicals having the characteristics of POPs are discussed in this book. In particular, since all the POPs are OCs, the terms POPs and OCs are used interchangeably throughout the text, and the readers should not be confused by this usage.

Effects on Wild Animals

The endocrine disrupting effects of POPs may be one of their major impacts on wildlife. Exposure to POPs through the environment and/or food chain, particularly during critical periods of life, may alter the immune system of birds and mammals; induce thyroid dysfunction in fish, shellfish and mammals; decrease fertility rates in some wild animals (e.g. mink) and may also cause disruptions in the sex characteristics of individual animals, thereby altering the sex ratio of the population (Colborn et al., 1996). Exposure to POPs has been linked to increased breast cancer and decreased semen quality in humans. Scores of animals with high POP concentrations have been found to suffer from tumours and skin lesions.

Recently, human activity is causing massive extinction of various animal species, forests and wildlife. POPs were also reported to affect wildlife in various ways. Myriad of reports suggest toxic impacts of POPs on plankton, mollusks, fish, birds and mammals (Tanabe et al., 1994; Muir et al., 1995; Farrington and Trip, 1995; Tanabe, 2000; Hanazato, 2001; Takeuchi et al., 2001; de Brito et al., 2002; Corsi et al., 2003; Dauwe et al., 2003; Kunisue et al., 2003). POPs have penetrated almost all the ecosystems and are now ubiquitous; this is evidenced by their detection in all environmental compartments and biota (Aono et al., 1997; Norstrom et al., 1998; Muir et al., 2000).

Environment Canada has stated in its report (Environment Canada, 1996) that POPs threaten animals on three levels—the genetic level, population level and community/ecosystem level. There are various impacts on the hormonal and enzyme systems of terrestrial and aquatic animals; these impacts lead to genetic dysfunctions at the sub-organismic level. Sexual dysfunction in many marine species of fish and mollusks and eggshell thinning in some bird species led to reduced reproductive success in individuals with high POP concentrations in tissues or organs (Reijnders et al., 1999; Konstantinou et al., 2000; Van der Oost et al., 2003).

Tanabe (2002) critically evaluated the ecotoxicological impact of persistent OCs on marine mammals. Colborn and Smolen (1996) listed 65 epidemiological abnormalities such as endocrine disruption, immunologic and reproductive dysfunction, tumours and decline in marine mammal populations, which have been observed since 1968. Simmonds (1991) stated that out of the 11 outbreaks of mass mortalities of marine mammals, 9 outbreaks have occurred after the 1970s around the coastal waters of developed and industrialized countries, further insinuating the adverse effects of toxic contaminants. All the above abnormalities may be attributed to the abnormal levels of any one or multitude of POPs.

Because of their persistence and bioaccumulative property, POPs and some other persistent organic contaminants are listed as PTS that pose a serious threat to biodiversity. Since the seas and oceans are the final recipients and reservoirs of POPs, animals living in these ecosystems, particularly the coastal ecosystems, are at a high health risk.

Biodiversity is always at its peak in tropical areas, and many of the countries in the tropical belt are known as mega-biodiversity nations. In addition, many bird species migrate from other parts of the globe to tropical countries, whereas many marine mammal species are known to migrate over short or long distances through coastal waters of tropical countries.

Diverse types of plants and animals boost ecosystem productivity in such mega-biodiversity nations. Each species plays an important role, and this enables the ecosystem to prevent and recover from a variety of disasters. A larger number of species translates into a diverse variety of plant and animal food for all organisms and hence the presence of a sustainable ecosystem. Therefore, it is imperative to save the biodiversity in tropical countries for maintaining a balance in the global ecosystem. Conservation of animal species requires knowledge on the contaminant effects. This has necessitated the studies on bioaccumulation, biomagnification, bioindicators and biomarkers.

Bioindicators—Concept and Definition

During the early days of monitoring of the marine environment, only easily measurable physicochemical variables such as atmospheric and water temperatures, salinity, dissolved oxygen, light penetration, algal blooms, nutrients such as nitrogen and phosphorus, heavy metals and organic carbon were monitored. In some cases, sediment characteristics were also examined. The data obtained were indirectly correlated with faunal and floral populations. Few actual measurements of the toxicant concentrations in the biotic components of the ecosystems were quantified.

In the late 1950s, when scientists realized the effects of bioaccumulation of OCs in animals, a momentum was gained to monitor the concentrations and the ensuing effects of chemicals in the tissues and organs of organisms. While bioindicator organisms were used to estimate the pollutant levels in an ambient environment by measuring the concentrations of pollutants in their bodies, biomarker responses were liberally used to measure both the levels and toxic effects of pollutants in the environment and biota.

Need for Bioindicator Studies

The advent of pesticide use in the early 1940s led to the development of the science of residue chemistry mainly because of concerns regarding human poisoning. Despite the current availability of sophisticated equipments and laboratory facilities, the recovery of residues from animal tissues is never 100%. The situation is worse when environmental samples are analyzed; besides the presence of ultra-low residue concentrations, natural substances present in the samples cause immense interference in analytical methodologies and quantification. During the early years of residue chemistry, multi-laboratory verification of residue concentrations in environmental samples seldom yielded similar results. In those days, chemists faced a new challenge of developing methods to recover and quantify the ultra-low concentrations of contaminants from environmental samples. Such analyzes were difficult and demanded complicated procedures. However, residue chemistry and field biology are not exact sciences, and since most of the results are based on assumptions and coincidence, interdisciplinary interpretation of results is difficult.

Before the Second World War, only approximately 30 inorganic and natural pesticides were in limited use (Moore, 1965). By 1945, many synthetic organic insecticides were being used, and the techniques for analysis of their residue concentrations were being developed (Woodard et al., 1945; Bryson et al., 1949; Carter et al., 1949). Wildlife biologists lagged by several years in their concern over exposure of animals to persistent pollutants, and some of the first investigations on pesticide effects were conducted with no support from residue data (Linduska and Surber, 1948).

The presence and persistence of pesticides in soil, water, air, plants and animals raised concerns regarding their long-term accumulation and effects. A program for monitoring their presence, persistence, transportation, bioaccumulation and biomagnification and toxicity became essential, and several programs were undertaken in earlier years (Moore, 1966, O'shea et al., 1980). However, the data obtained from environmental samples did not represent the actual scenario at the sampling sites, particularly in aquatic habitats, because of the highly varying conditions of the environmental parameters. This necessitated the use of animal sentinels in such highly mobile environments.

Whether analyzing residues in mollusks, seabird eggs, fish, birds or mammalian tissues, most biologists have their own concepts regarding the best indicator species (Elliott et al., 1988; Farrington and Trip, 1995; Minh et al., 2002; Tanabe, 2002; Kunisue et al., 2003; Monirith et al., 2003; Ueno et al., 2003). Furthermore, small mammals, as a group, were proposed as biomonitors of contamination (Talmage and Walton, 1991) and animals, in general, for monitoring environmental quality (Buck, 1979) and as indicators of pollution (Bruggemann et al., 1974). Such monitoring programs produced information that reinforced the conclusions gained from other investigations and underlined the importance of the use of biological indicators in monitoring environmental pollution by persistent chemicals.

What are Bioindicators?

There are several approaches to measure chemical contaminants in the marine environment. For many chemicals, monitoring by analytical chemistry has furnished an extensive database of the concentrations of contaminants in various species and sites. The utility of this approach has been proved by the documentary evidence on geographic and temporal differences in coastal pollution collected through this approach. 'Mussel Watch' and 'Status and Trend Monitoring Programs' are two appropriate examples of such biomonitoring programs in which the concentrations of a variety of pollutants in bivalve mollusks were measured and utilized for explaining the differences in pollutant levels with respect to locations and time trends (Goldberg, 1975; Phillips, 1980, 1985; Farrington et al., 1983, 1987; Farrington and Trip, 1995; Monirith et al., 2003). The National Contaminant Biomonitoring Program (NCBP) has highlighted the possibility of using fish as suitable bioindicators for monitoring OCs in the aquatic environment (Schmidt et al., 1990).

The processing of environmental samples (e.g. water and air) is tedious because these contain ultra-low POP concentrations, which exhibit an intrinsic hydrophobic character. Therefore, an alternative way of measuring the environmental contaminants is determining their concentrations in the animals that are directly exposed to these chemicals through their ambient environment (Phillips, 1995). The POPs concentrations measured in an indicator organism can be used to estimate their concentration in the surrounding environment. Such organisms have been referred to as 'bioindicators'. Such measurements have enabled environmental managers to extrapolate the results obtained from wildlife to humans.

Alternatives to such chemical measurements of pollutants, which are often not cost-effective or insensitive, include indirect techniques such as immunoassays using chemical-specific antibodies (Szurdoki et al., 1996). Assays employing such biomarkers offer another powerful alternative to chemical assays (Lam and Gray, 2003). These methods involve the quantitative measurement of changes occurring in the biological system in response to exposure to xenobiotic substances.

Bioindicators are typically used to measure the actual concentrations of pollutants in the biological systems at higher levels of organization, while biomarkers are generally used to indicate the exposure of organisms to contaminants at lower levels of biological organization. There are many definitions of the terms 'biomarkers' and 'bioindicators'. Until the early 1990s, little effort was directed towards differentiating bioindicators and biomarkers. All the observations made using animal sentinels were referred to as bioindications. For example, in 1987, the National Research Council (NRC) of the National Academy of Sciences defined a bioindicator as a measure of organismic fluid, tissue, cell or other biotic factors that indicates the presence and magnitude of stress response (NRC, 1987). In contrast, the Environmental Monitoring and Assessment Program (EMAP) prefers to use the characteristics of the environment to quantify the magnitude of stress, degree of exposure to stressor and degree of ecological response to stress (Hunsaker and Carpenter, 1990). The Commission on Life Sciences defined the animal sentinel systems (bioindicators) as systems in which data on animals exposed to environmental contaminants are regularly and systematically collected and analyzed to identify potential health hazards to other animals or humans (CLS, 1991).

Along with the term bioindicators, the term biomarker has been gaining acceptance in recent years, albeit with some inconsistency in definition (Walker et al., 2001). The authors defined biomarkers as 'any biological response to an environmental chemical at or below the individual level, demonstrating a departure from the normal status'.

Bioindicators can be defined as the animal(s) or plant(s) that accumulates pollutants in its tissues and organs in direct proportion to the environmental levels. In this book, we consider the bioaccumulation of chemicals that occurs at and above the organismic (organs and tissues) level as a bioindicator response and the organisms that can biomagnify the chemicals to measurable levels in direct proportion to the changes in the ambient environment as bioindicators. Changes that occur at the population and assemblage levels are more often referred to as 'bioindicators' (hence, sometimes, the term 'bioindicator organisms' is used); however, this need not be considered for the purpose of this book.

It is argued that chemistry-based monitoring programs cannot provide information on biological effects. When the presence and levels of contaminants (e.g. the presence of any one or more POPs) need to be determined, the measurement of its concentrations in a suitable bioindicator organism will be an appropriate approach. Biomonitoring data are useful for ecological risk assessment. Biomarker analyzes usually provide the information of the occurrence of an effect. Biomonitoring data can be used to quantitatively assess the risks by the top-down or bottom-up approach. As stated by Lam and Gray (2001), chemistry data will also be useful in predicting potential biological effects if prior biomarker investigations have been successful in relating contaminant levels in the environmental samples with responses in biological systems. Under this circumstance, the use of chemical data to predict contaminant effects may be cost-effective.

In places known to be hot spots of pollutants, measurement of biomarker responses on a dose-response basis will be more effective and sensitive. Bioindicators are the measures of the actual chemistry-based concentrations in the organisms resulting from long-term exposures, whereas biomarkers provide an early warning signal of contamination in the environment although they are not contaminant specific. In reality, both bioindicator and biomarker approaches have their respective merits and demerits, but they can always compliment each other.

Essential Characteristics

This book aims to recommend bioindicators suitable for measuring the POPs concentrations and extrapolating these data to environmental contamination; in addition, the book suggests possible human impacts of POPs in developing countries. This immediate need has already been stressed in previous GEF reports. Therefore, in this book, we will be restricting ourselves to the following points.

1. Explaining the essential characteristics of a bioindicator organism for evaluation of POPs in the marine environment
2. Evaluating the available literature on the measurement of POPs in various animal species
3. Determining the occurrence and concentrations of POPs in various tissues, organs and bodies of different animals
4. Investigating the relationship between the POP concentrations in animal tissues and those in the environment
5. Identifying the spatial and temporal variations in the POP concentrations in animal tissues and organs
6. Selecting suitable organs and tissues for the evaluation of different POPs
7. Suggesting suitable bioindicator organisms that can be used in developing countries for the evaluation of the 12 POPs, stipulated by the GEF for immediate action

The biological effects of suspected toxic substances in animals can be evaluated while the animals remain in their natural habitats. This may offer an opportunity to assess the intensity of exposure and measure the effects of chemical mixtures and low-level exposure over longer periods. Animals can be normally placed in an area of special interest, that is, an area of known contamination to determine the extent or temporal changes in contamination. Before the type of program and nature of the animal sentinel (bioindicator organism) are chosen, several characteristics need to be taken into consideration. In other words, the essential characteristics of the indicator organism should be decided.

Various animal species can be potentially used (or have been used) as bioindicator organisms for measuring diverse pollutants under different situations. Goldberg (1975), Phillips (1980), CLS (1991), Summers et al. (1996), Adams and Greeley (2000) and Lam and Gray (2003) suggested that several attributes of an animal decide its suitability as a bioindicator. To use animal bioindicators for measuring concentrations and effects of POPs in developing countries, we consider the following characteristics to be essential.

1. *A bioindicator organism for a POP should elicit a measurable response, including accumulation of that POP in any particular organ or tissue or throughout its body*: The animal should be able to accumulate the chemical in the whole body or in a particular organ or tissue to several times its concentration in the ambient environment, thereby enabling easy quantification with less sophisticated instruments, which is a prerequisite for developing countries. At the same time, if the animal is less sensitive to high doses of the chemical, specimens can be collected at intervals and tested for body burdens or even monitored for long-term dose response.

2. *The species should have a territory that overlaps the area to be monitored*: If the sentinel animal used for monitoring has a territory within or beyond the scope of the intended study area, the animal would be inappropriate for use as a bioindicator. Normally, migratory animals should be avoided. An organism living within the study area throughout its life cycle may be a good bioindicator organism.

3. *The specimens should be easy to collect*: The animals should inhabit an environment, either terrestrial or aquatic, from where they can be collected easily.

4. *The handling, dissection and sampling of the target tissues and organs should be easy*: If a large sampling is required, the specimen should be of a reasonable size and should be easily dissectible by any team member at the collection site.

5. *The specimens should be available in adequate numbers*: Rare or endangered species might pose several problems because of the difficulty involved in locating them. They may also obscure the results because of the

prevailing population stress. For measuring the time trends and seasonal trends, many specimens may be required for statistical analyzes of the data over a period of time; this may necessitate a larger sample size. Therefore, in general, the population of a bioindicator species should be sufficiently large to sustain the harvesting required for the monitoring study without eliciting any adverse effects.

6. *The species should have an optimum lifespan that is sufficiently long for substantial accumulation or adverse changes to occur*: Animals with a very short or very long lifespan may not be good bioindicators. In the former case, the animal may not accumulate enough concentrations of the chemical for easy quantification. In the latter case, the chemical may plateau in the tissues and organs of the animal. Both cases will not provide information regarding the time of exposure.

7. *The species should have a simple feeding habit*: Interpretation of data based on feeding habits may be difficult in animals with omnivorous feeding habits and in opportunistic feeders.

8. *The species should be preferably sedentary or have a very short migration within the study area*: Sedentary animals, for obvious reasons, serve as good bioindicator species for measuring the pollution status in discrete locations. In case of non-availability of a sedentary species (in a terrestrial ecosystem), an animal species with a small home range should be collected.

9. *The species should withstand a wide range of environmental and climatic conditions*: This characteristic is important for measuring the seasonal variations in the chemical concentrations. Moreover, the stress occurring due to environmental changes should not be reflected in the accumulation pattern of the chemical(s) in the animal body.

10. *The species should be taxonomically well known and stable*: The researcher should be able to identify the species clearly and distinguish it from related species.

11. *The biology and general life history should be well known*: Knowledge regarding the life history of the species is also essential because several life history parameters such as feeding, migration, egg laying, parturition, lactation and hibernation drastically affect the POP concentrations in animals regardless of their lifespans.

12. *Survey and marking of individuals of a population should be easy*: In case of long-term field studies, the researcher should be able to locate and observe the subjects easily.

13. *The species should occupy a broad geographical range*: In a global survey, it is extremely important that a species occupies a broad geographical range. The species as a whole should inhabit several parts of the world,

not just restricted areas. Thus, the data from different parts of the world can be accurately compared.

14. *Accumulation patterns observed in a target bioindicator species should be reflected at least in some other related and unrelated species co-inhabiting the area*: Data obtained from the bioindicator species should be applicable or extrapolatable to other species in the ambient ecosystem. Thereby, the species will be an efficient bioindicator of the ecosystem. This characteristic may be useful because when the species is not available during a particular sampling period, a related species from the ecosystem can be collected.

15. *The animal should be of commercial and economic importance in the area*: It is crucial to prove the economic importance of a species under study. This is particularly important in countries where resources are scarce and cost-efficacy of a program is the decisive factor.

16. *The collection or purchase of specimens should be cost-effective*: This may be important for the program in which many developing countries have to undertake their own programs based on the GEF recommendations.

17. *The specimens should be internationally transportable without any legal impediment*: Since the specimens have to be transported to regional laboratories within a country or abroad, this condition is fully applicable to the specimens to be used as bioindicators in developing countries.

The above stated criteria may vary in importance depending on the purpose of the study. However, in general, the species that can fulfil most of the above criteria may be recommended for further studies for monitoring POPs in developing countries. Based on the above criteria, we will discuss the studies thus far carried out on different taxa of animal species and evaluate their suitability for our purpose, in the forthcoming chapters.

References

Adams, S.M. and M.S. Greeley, 2000. Ecotoxicological indicators of water quality: using multi-response indicators to assess the health of aquatic ecosystems. Water, Air, and Soil Pollut., 123: 103–115.

Addison, R.F. and T.G. Smith, 1974. Organochlorine residue levels in Arctic ringed seals: variation with age and sex. Oikos, 25: 335-337.

AMAP, 1998. S.J. Wilson, J.L. Murray and H.P. Huntington (Eds.), Assessment Report: Arctic Pollution Issues. Arctic Monitoring and Assessment program, p. 859.

Aono, S., S. Tanabe, Y. Fujise, H. Kato and R. Tatsukawa, 1997. Persistent organochlorines in minke whale (*Balaenoptera acutrostrata*) and their prey species from the Antarctic and the North Pacific. Environ. Pollut., 98: 81–89.

Barakat, A.O., 2004. Assessment of persistent toxic substances in the environment of Egypt. Environ. Int., 30: 309–322.

Barnett, D.C., 1950. The effect of some insecticide sprays on wildlife. Proc. Annu. Conf. West. Assoc. State Game Fish Comm., 30: 125–134.

Beyer, W.N., G.H. Heinz and A.W. Redmon-Norwood, 1996. Environmental contaminants in wildlife. Interpreting tissue concentrations. Lewis Press, Boca Raton, London, New York, Tokyo, p. 494.

Bruggemann, J., L. Busch, U. Drescher-Kaden, W. Eisele and P. Hoppe, 1974. Pesticide residues in organs of wild living animals as indicator of pollution. Int. Congr. Game Biol., 11: 439–449.

Bryson, M.J., C.I. Draper, J.R. Harris, C. Biddulph, D.A. Greenwood, L.E. Harris, W. Binns, M.L. Miner and L.L. Madsen, 1949. DDT in eggs and tissues of chickens fed varying levels of DDT. Proc. Am. Soc. Hortic. Sci., 54: 232–236.

Buck, W.B., 1979. Animals as monitors of environmental quality. Vet. Hum. Toxicol., 21: 277–284.

Carter, R.H., R.W. Wells, R.D. Redcleff, C.L. Smith, P.E. Hubanks and H.D. Mann, 1949. The chlorinated hydrocarbon content of milk from cattle sprayed for control of horn flies. J. Econ. Entomol., 42: 116–118.

CLS, 1991. Commission on Life Sciences - Animals as Sentinels of Environmental Health Hazards. The National Academic Press, p. 160.

Cohen, M., B. Commoner, A.E. Cabrera, D. Muir and C.S. Burgoa, 1997. Dioxin: A case study. Draft Report to the Secretariat of the Commission on Environmental Cooperation, Montreal.

Colborn, T. and M.J. Smolen, 1996. Epidemiological analysis of persistent organochlorine contaminants in cetaceans. Rev. Environ. Contam. Toxicol., 146: 91–172.

Colborn, T., D. Dumanoski and J.P. Myers, 1996. Our Stolen Future. Dutton, New York, USA, p. 306.

Corsi, I., M. Mariottini, C. Sensini, L. Lancini and V.delle Cerchaia, 2003. Fish as bioindicators of brackish ecosystem health: integrating biomarker responses and target pollutant concentrations. Oceanol. Acta, 26: 129–138.

Dauwe, T., S.G. Chu, A. Covaci, P. Schepens and M. Eens, 2003. Great tit (*Parus minor*) nestlings as biomonitors of organochlorine pollution. Arch. Environ. Contam. Toxicol., 44: 89–96.

De Witt, J.B., J.V. Derby Jr. and F. Mangan Jr., 1955. DDT vs. wildlife. Relationships between quantities ingested, toxic effects and tissue storage. J. Am. Pharm. Assoc. Sci. Ed., 44: 22–24.

de Brito, A.P.X., S. Takahashi, D. Ueno, H. Iwata, S. Tanabe and T. Kubodera, 2002. Organochlorine and butyltin residues in deep-sea organisms collected from the western North Pacific, off-Tohoku, Japan. Mar. Pollut. Bull., 45: 348–361.

Edwards, C.A., 1973. Environmental Pollution by Pesticides. Plenum Press, New York, p. 542.

Elliott, J.E., R.J. Norstrom and J.A. Keith, 1988. Organochlorines and egg shell thinning in northern gannets (Sula bassanus) from eastern Canada, 1968–1984. Environ. Pollut., 52: 81–102.

Environment Canada, 1996. Proposed Interim Canadian Sediment Quality Guidelines, Ottawa. Guidelines Division, Science Policy and Environmental Quality Branch, Ecosystem Science Directorate, Environmental Conservation.

Farrington, J.W. and B.W. Tripp, 1995. International Mussel Watch Project. Initial Implementation Phase. Final Report. NOAA, U.S. Department of Commerce, p. 63.

Farrington, J.W., E.D. Goldberg, R.W. Risebrough and J.H. Martin, 1983. U.S. "Mussel Watch" 1976–1978: An overview of the trace metal, DDE, PCB, hydrocarbon and radionuclide data. Environ. Sci. Technol., 17: 490–496.

Farrington, J.W., A.C. Davies, B.W. Tripp, D.K. Phelps and W.B. Galloway, 1987. 'Mussel Watch'—Measurements of chemical pollutants in bivalves as one indicator of coastal environmental quality. In: T.P. Boyle (Ed.), New Approaches for Monitoring Aquatic Ecosystems, ASTM STP 940, American Society of Testing Materials, Philadelphia, pp. 125–139.

Gannon, N. and J.H. Bigger, 1958. The conversion of aldrin and heptachlor to their epoxides in soil. J. Econ. Entomol., 51: 1–2.

Gannon, N. and G.C. Decker, 1959. Insecticide residues as hazards to warm-blooded animals. Trans. N. Am. Wildl. Conf., 24: 124–132.

Goldberg, E.D., 1975. The mussel watch-A first step in global marine monitoring. Mar. Pollut. Bull., 6: 111.

Hanazato, T., 2001. Pesticide effects on freshwater zooplankton: an ecological perspective. Environ. Pollut., 112: 1–10.

Hunsaker, C.N. and D.E. Carpenter, 1990. Environmental Monitoring and Assessment Program - Ecological Indicators, EPA/600/3-060.

Hunt, E.G. and J.D. Linn, 1970. Fish kills by pesticides. In: J.W. Gillett (Ed.), The Biological Impacts of Pesticides in the Environment. Environ. Health Ser. No. 1, Oregon State University, Corvallis, Oregon, pp. 97–102.

Iwata, H., S. Tanabe, N. Sakai and R. Tatsukawa, 1993. Distribution of persistent organochlorines in the oceanic air and surface seawater and the role of ocean on their global transport and fate. Environ. Sci. Tech., 27: 1080–1098.

Iwata, H., S. Tanabe, N. Sakai, A. Nishimura and R. Tatsukawa, 1994.

Geographical distribution of persistent organochlorines in air, water and sediments from Asia and Oceania, and their implications for global redistribution from lower latitudes. Environ. Pollut., 85: 15–33.

Jones, K.C. and P. de Voogt, 1999. Persistent organic pollutants (POPs): state of the science. Environ. Pollut., 100: 209–221.

Konstantinou, I.K., V. Goutner and T.A. Albanis, 2000. The incidence of polychlorinated biphenyl and organochlorine pesticide residues in eggs of the cormorant (*Phalacrocorax carbo sinensis*): an evaluation of the situation in four Greek wetlands of international importance. Sci. Total Environ., 257: 61–79.

Kunisue, T., M. Watanabe, A.N. Subramanian, A. Sethuraman, A. Titenko, V. Qui, M. Prudente and S. Tanabe, 2003. Accumulation features of persistent organochlorines in resident and migratory birds from Asia. Environ. Pollut., 125: 157–172.

Lam, P.K.S. and J.S. Gray, 2001. Predicting effects of toxic chemicals in the marine environment. Mar. Pollut. Bull., 42: 169–173.

Lam, P.K.S. and J.S. Gray, 2003. The use of biomarkers in environmental monitoring programs. Mar. Pollut. Bull., 46: 182–186.

Linduska, J.P. and E.W. Surber, 1948. Effects of DDT and other insecticides of fish and wildlife. Summary of investigations during 1947, U.S. Dep. Inter. Fish Wildl. Serv. Circ., 15. p. 19.

Loganathan, B.G. and K. Kannan, 1994. Global organochlorine contamination trends: an overview. Ambio, 23: 187–191.

Minh, T.B., T. Kunisue, N.T.H. Yen, M. Watanabe, S. Tanabe, N.D. Hue and V. Qui, 2002. Persistent organochlorine residues and their bioaccumulation profiles in resident and migratory birds from north Vietnam. Environ. Toxicol. Chem., 21: 2108–2118.

Minh, N.H., T.B. Minh, M. Watanabe, T. Kunisue, I. Monirith, S. Tanabe, S. Sakai, A.N. Subramanian, K. Sasikumar, P.H. Viet, B.C. Tuyen, T.S. Tana and M.S. Prudente, 2003. Open dumping site in Asian developing countries: a potential source of polychlorinated dibenzo-*p*-dioxins and polychlorinated dibenzofurans. Environ. Sci. Tech., 37: 1493–1502.

Mohr, R.W., H.S. Telford, E.H. Peterson and K.C. Walker, 1951. Toxicity of orchard insecticides to birds. Wash. Agri. Exp. St., Circ. No. 170, p. 22.

Monirith, I., D. Ueno, S. Takahashi, H. Nakata, A. Sudaryanto, A.N. Subramanian, S. Karuppiah, A. Ismail, M. Muchtar, J. Zheng, B.J. Richardson, M. Prudente, N.D. Hue, T.S. Tana, A.V. Tkalin and S. Tanabe, 2003. Asia-Pacific mussel watch: monitoring contamination of persistent organochlorine compounds in coastal waters of Asian countries. Mar. Pollut. Bull., 46: 281–300.

Moore, N.W., 1965. Pesticides and birds - a review of the situation in Great Britain in 1965. Bird Study, 12: 222–252.

Moore, N.W., 1966. A pesticide monitoring system with special reference to the selection of indicator species. J. Appl. Ecol., 3: 261–269.

Muir, D.C.G., M.D. Segstro, K.A. Hobson, C.A. Ford, R.E.A. Stewart and S. Olpinski, 1995. Can elevated levels of PCBs and organochlorine pesticides in walrus blubber from eastern Hudson Bay (Canada) be explained on consumption of seals? Environ. Pollut., 90: 335–348.

Muir, D., F. Riget, M. Clemann, J. Skaare, L. Kleivane, H. Nakata, R. Dietz, T. Severinsen and S. Tanabe, 2000. Circumpolar trends of PCBs and organochlorine pesticides in the Arctic marine environment inferred from ringed seals. Environ. Sci. Technol., 34: 2431–2438.

Nakata, H., S. Tanabe, R. Tatsukawa, Y. Koyama, N. Miyazaki, S. Belikov and A. Boltunov, 1998. Persistent organochlorine contaminations in ringed seals (*Phoca hispida*) from the Kara Sea, Russ. Arctic. Environ. Toxicol. Chem., 17: 1745–1755.

Norstrom, R.J., M. Simon, D.C.G. Muir and R.E. Schweinsburg, 1998. Organochlorine contaminants in Arctic marine food chains: Identification, geographical distribution and temporal trends in polar bears. Environ. Sci. Technol., 22: 1063–1071.

NRC, 1987. Committee on Biological Markers of the National Research Council, 1987. Biological markers in environmental health research. Environ. Health Perspect., 74: 3–9.

Orris, P., L.K. Chary, K. Perry and J. Ausbury, 2000. Persistent Organic Pollutants and Human Health. A publication of the World Federation of Public Health Association's Persistent Organic Pollutants Project. p. 38.

O'shea, T.J., R.L. Brownell Jr., D.R. Clark Jr., W.A. Walker, M.L. Gay and T.G. Lamont, 1980. Organochlorine pollutants in small cetaceans from the Pacific and South Atlantic Oceans. November 1968–June 1976. Pestic. Monit. J., 14: 35–46.

Parr, T.W., A.R.J. Sier, R.W. Battarbee, A. Mackay and J. Burgess, 2003. Detecting environmental change: science and society—perspectives on long-term research monitoring in the 21st century. Sci. Total Environ., 310: 1–8.

Phillips, D.J.H., 1980. Quantitative Aquatic Biological Indicators. Their Use to Monitor Trace Metal and Organochlorine Pollution. Applied Science Publishers, London, p. 488.

Phillips, D.J.H., 1985. Organochlorines and trace metals in green-lipped mussels Perna viridis from Hong Kong waters: a test for indicator availability. Mar. Ecol. Prog. Ser., 21: 251–258.

Phillips, D.J.H., 1995. The chemistries and environmental fates of trace metals and organochlorines in aquatic ecosystems. Mar. Pollut. Bull., 31: 193–200.

Post, G., 1952. The effects of aldrin on birds. J. Wildl. Manage., 16: 492–497.

Ramesh, A., S. Tanabe, H. Iwata, R. Tatsukawa, A.N.D. Mohan and V.K. Venugopalan, 1990. Seasonal variation of persistent organochlorine insecticide residues in Vellar river waters in Tamil Nadu, South India. Environ. Pollut., 67: 289–304.

Reijnders, P.J.H., G.P. Donovan, A. Aguilar and A. Bjorge, 1999. (Eds.), Report of the workshop on chemical pollution and cetaceans. J. Cetacean Res. Manage., Spl. Issue, p. 53.

Risebrough, R.W., P. Reiche, S.G. Herman, D.B. Peakall and M.N. Kirven, 1968. Polychlorinated biphenyls in the global ecosystem. Nature (Lond.), 220: 1098–1102.

Ritter, L., K.R. Solomon and J. Forget, 1995. An Assessment Report on: DDT-Aldrin-Dieldrin-Endrin-Chlordane-Heptachlor-Hexachlotoben-zene-Mirex-Toxaphene-Poly-chlorinated Biphenyls-Dioxins and Furans. For: The International program on Chemical Safety (IPCS) within the framework of the Inter-Organization program for the Sound Management of Chemicals (IOMC). p. 24.

Rolland, R., M. Gilbertson and T. Colborn, 1995. Environmentally induced alterations in development: a focus on wildlife. Environ. Health Perspectives, 103: 3–5.

Rosene, W. Jr., 1965. Effects of field applications of heptachlor on bobwhite quail and other wild animals. J. Wildl. Manage., 29: 554–580.

Schmidt, C.J., J.L. Zajicek and P.H. Peterman, 1990. National contaminant bio-monitoring program: Residues of organochlorines in U.S. freshwater fish, 1976–1984. Arch. Environ. Contam. Toxicol., 19: 748–781.

Simmonds, M., 1991. Marine mammal epizootics worldwide. In: X. Poster and M. Simmonds (Eds.), Proceeding of Mediterranean Striped Dolphins Mortality International Workshop, Green Peace International Mediterranean Sea Project, Madrid, Spain, pp. 9–19.

Sladen, W.J., C.M. Menzie and W.L. Reichel, 1966. DDT residues in Adelie penguins and a crab eater seal from Antarctica: ecological implications. Nature (Lond.), 210: 670–673.

Stickel, L.F., 1973. Pesticide residues in birds and mammals. In: C.A. Edwards (Ed.), Environmental Pollution by Pesticides. Plenum Press, New York, pp. 254–312.

Subramanian, B.R., S. Tanabe, H. Hidaka and R. Tatsukawa, 1983. DDTs and PCB isomers and congeners in Antarctic fish. Arch. Environ. Contam. Toxicol., 12: 621–626.

Summers, R., L. Folmar and M. Rondon-Naveira, 1997. Development and testing of bioindicators for monitoring the condition of estuarine ecosystems. Environ. Monit. Assess., 47: 275–301.

Szurdoki, F., L. Jaeger, A. Harris, H. Kido, M. Goodrow, A. Szkacs, M. Wothrberg, J. Zheng, D. Stoutamire, J. Sanborn, S. Gilman, A. Jones, P. Choudry and B. Hammock, 1996. Rapid assays for environmental and biological monitoring. J. Environ. Sci. Health Pt. B, 31: 451–458.

Takeoka, H., A. Ramesh, H. Iwata, S. Tanabe, A.N. Subramanian, D. Mohan, A. Magendran and R. Tatsukawa, 1991. Fate of the insecticide HCH in the tropical coastal areas of South India. Mar. Pollut. Bull., 22: 290–297.

Takeuchi, I., S. Takahashi, S. Tanabe and N. Miyazaki, 2001. Caprella watch: a new approach for monitoring butyltin residues in the ocean. Mar. Environ. Res., 52: 97–113.

Talmage, S.S. and B.T. Walton, 1991. Small mammals as monitors of environmental contaminants. Rev. Environ. Contam. Toxicol., 119: 47–145.

Tanabe, S., 2000. Asian developing regions: persistent organic pollutants in the seas. In: Sheppard, C.R.C. (Ed.), Seas at the Millennium: An Environmental Evaluation, Pergamon, Amsterdam, The Netherlands, pp. 447–462.

Tanabe, S., 2002. Contamination and toxic effects of persistent endocrine disrupters in marine mammals and birds. Mar. Pollut. Bull., 45: 69–77.

Tanabe, S., S. Miura and R. Tatsukawa, 1986. Variations of organochlorine residues with age and sex in Antarctic minke whale. Mem. Natl. Inst. Polar Res., 44: 174–181.

Tanabe, S., H. Iwata and R. Tatsukawa, 1994. Global contamination by persistent organochlorines and their ecotoxicological impact on marine mammals. Sci. Total Environ., 154: 163–177.

Ueno, D., T. Inoue, K. Ikeda, H. Tanaka, H. Yamada and S. Tanabe, 2003. Specific accumulation of polychlorinated biphenyls and organochlorine pesticides in Japanese common squid as a bioindicator. Environ. Pollut., 125: 227–235.

Ueno, D., N. Kajiwara, H. Tanaka, A.N. Subramanian, G. Fillmann, P.K.S. Lam, G.J. Zheng, M. Muchitar, H. Razak, M. Prudente, K. Chung and S. Tanabe, 2004. Global distribution of polychlorinated biphenyl ethers using skipjack tuna as a bioindicator. Environ. Sci. Technol., 38: 2312–2316.

UNEP Chemicals, 2003. Regionally Based Assessment of Persistent Toxic Substances. Global Report 2003, p. 207, http: //www.chem.unep.ch/pts.

Van der Oost, R., J. Beyer and N.P.E. Vermeulen, 2003. Fish bioaccumulation

in environmental risk assessment: a review. Environ. Toxicol. Pharmacol.,
13: 57–149.

Walker, C.H., S.P. Hopkin, R.M. Sibly and D.B. Peakall, 2001. Principles of
Ecotoxicology. Second Edition. Taylor and Francis, New York, p. 309.

Wania, F. and D. Mackay, 1993. Global fractionation and cold condensation of
volatile organochlorine compounds in polar regions. Ambio, 22: 10–18.

Woodard, G., R.R. Ofner and C.M. Montgomery, 1945. Accumulation of DDT in
the body fat and its appearance in the milk of dogs. Science (Washington,
D.C.), 102: 177–178.

WWF, 1999. Beneficial bugs at risk from pesticides. www.neteffct.ca/pesticides/
resources/bugs-at.risk.pdf

Mussels: Universal Pollution Indicators

Bivalve mollusks have been extensively used and proved successful as bioindicators for monitoring POPs in natural waters. These organisms, particularly mussels and oysters, possess all the characteristics required of a bioindicator for POPs. Brackish water species belonging to the genera *Perna*, *Mytilus* and *Crassostrea* were found to be most suitable. Freshwater mussels can be used for monitoring inland areas.

Chapter 2: Mussels

Introduction

Growth in industrial activity during the 20th century has resulted in a rapid increase in the release of chemicals, either mobilized or synthesized by man, into estuarine and coastal environments. Many of these chemicals are bioaccumulated within the tissues of biota to concentrations above ambient levels in the environment. As already discussed earlier in this review, different organisms such as fish, birds and mammals have been successfully used as bioindicators of environmental pollution. Such bioindicators provide integrated information regarding environmental contamination and effects that cannot be determined by the chemical analyses of water samples. The ultimate aim of environmental toxicology is to predict and diagnose the biological effects resulting from exposure to chemicals in the environment. To meet this objective, it is necessary to first establish the relationship between the chemical contaminants in the environment and those present in the tissues of biota.

Monitoring of trace toxic substances in the aquatic environment by using biological indicators is a well-established method. Specific programs for monitoring pesticide residues were undertaken as early as the 1960s by using various animals such as birds, mollusks and mammals (Dewitt et al., 1960; Butler, 1966; Moore, 1966). Johnson et al. (1967) described the monitoring of pesticide residues in fish and other wildlife by the US Fish and Wildlife Service. Cain (1981) published reports on the measurement of pesticides in birds. Buck (1979) proposed animals, in general, as 'monitors of environmental quality'. Talmage and Walton (1991) suggested small mammals, as a group, to be 'monitors of environmental contaminants'. These authors had specific ideas regarding which animal(s) was the best indicator species, and they have proposed their own criteria for bioindicator species. Several monitoring programs produced enormous volumes of information reinforcing or opposing the conclusions drawn in previous investigations.

Mussel Watch

Experience gained in the 1960s and the early 1970s has led to the conclusion that the geographical extent of the severity of marine environmental contamination and the associated biological impact are largely unknown and undocumented.

Professor Goldberg of the Scripps Institution of Oceanography appealed for a global marine monitoring program to serve as a springboard for action (Goldberg, 1975). He proposed the establishment of a monitoring program named 'Mussel Watch' to assess the spatial and temporal trends in chemical contamination in estuarine and coastal areas. In his editorial in the journal Marine Pollution Bulletin, Prof. Goldberg has outlined a monitoring program to be conducted on a global scale. The program is based on the 'sentinel organism concept' that enables the detection of trends in the concentrations of several marine contaminants.

Since the mid-1970s, scientists from several countries have used filter-feeding bivalve mollusks to monitor selected contaminants in coastal marine waters. This has then led to the establishment of similar local or regional 'Mussel Watch' programs in many countries (e.g. UK, Japan, USA, France, Canada, Australia, Taiwan, India, Mediterranean countries, South Africa and the former USSR). The mussel watch program was initially used for analyzing trace metals and radionuclides and was then extended to several other pollutants such as pesticides and other POPs, alkyltins and hydrocarbons (Farrington et al., 1987).

The 1990s saw a rapid rise in the use of bioindicators as tools for measuring POP loads in the environment; however, there was simultaneous recognition of the fact that the data being collected are inadequate to meet the needs of the scientists and policy makers for streamlining pollution control. Thereafter, policy makers, scientists and the general public raised increasing concerns regarding gaining easy and rapid access to information. The best approach for rapid and easy studies on the environmental contamination in an ecosystem is to select a suitable bioindicator.

Why are Mussels Preferred as Bioindicators?
Mussels or other bivalves are commonly preferred for biomonitoring because they possess many advantages over other organisms. Several attributes make mussels superior to other organisms for use as 'sentinel' or 'indicator' organisms in worldwide environmental monitoring programs in both marine and estuarine environments (Phillips, 1980; Farrington et al., 1987; Tanabe, 2000). Phillips (1980), Gosling (1992) and Farrington and Trip (1995) have explained many advantages of using mollusks, particularly bivalves, as bioindicators of contaminant loads in coastal and estuarine ecosystems. It may be appropriate to mention some of these advantages here.
1. Bivalve species such as mussels and oysters have a wide geographical distribution and are dominant members of estuarine and coastal communities. Since bivalves of the same species can be collected from wide geographic locations, the problems associated with comparing data

obtained from different species are eliminated. This will be an important parameter particularly in tropical areas having a wide biodiversity.

2. Mollusks are sedentary and are therefore better than mobile species such as finfish as integrators of chemical contamination in a given area.

3. They are relatively tolerant to a wide range of environmental conditions such as salinity, season, sampling position in the water column, size and reproductive condition. Since these animals are sedentary, most of the problems that may arise due to these variables can be eliminated with relative ease during the sampling procedure (Phillips, 1980).

4. Bivalves are relatively tolerant to a wide range of environmental contaminants, including moderately high levels of many types of contaminants. They can exist in habitats simultaneously contaminated by a variety of pollutants.

5. A correlation always exists between the pollutant content in the organism and the average pollutant concentration in the surrounding habitat. POPs are usually bioconcentrated many-fold in bivalve mollusks.

6. Bivalves such as mussels and oysters always occur in wide and stable populations; hence, they can be sampled repeatedly during different seasons.

7. Many bivalves have a reasonably long lifespan (e.g. 1–8 years); therefore, specimens of various sizes (year-classes) can be sampled easily for comparison.

8. Most bivalves are reasonably sized; this provides adequate tissue for analysis.

9. Bivalves are suspension feeders (filter feeders) that pump several litres of water through their gills every hour and concentrate many chemicals in their tissues, by factors of $10–10^5$ relative to the concentrations in water. This enables easy measurement of contaminants.

10. In comparison with many other animals at the same trophic level, bivalves have a very low level of enzymatic activity for drug metabolism. Therefore, the contaminant concentrations in the tissues of bivalves more accurately reflect the magnitude of environmental contamination. At the same time, unlike in larger animals such as marine mammals and birds, which are also used as bioindicators, bioaccumulation in mussels adequately reflects the changing contaminant levels in the environment (Phillips, 1980; Farrington et al., 1987; Cossa, 1988).

11. Using pooled literature data on the uptake of contaminants by mollusks, Livingstone (1991, 1992) has shown that the rates of uptake of hydrocarbons generally exceed their rates of metabolism by an order of magnitude or more. This explains the marked bioaccumulation of these

compounds by mollusks, and the process can thus be explained based on a simple lipid/water equilibrium model (Burns and Smith, 1981).

12. Other factors that undoubtedly contribute to the long residence times of organic xenobiotics in mollusks and other marine invertebrates are the slow release of metabolites into seawater and the covalent binding of xenobiotics to macromolecules (Livingstone, 1991, 1992).

13. The measurement of chemicals in bivalve tissues enables the assessment of biological availability, which is not apparent from the measurement of contaminants in environmental compartments (water, suspended matter and sediment).

14. Bivalves are sufficiently sturdy to survive laboratory and field studies in cages.

15. Most bivalves are of commercial interest, and the measurement of chemical contaminants in bivalves is a matter of public interest.

Factors Affecting Bioaccumulation

Lipid Levels and Composition
Generally, when organic contaminant fluxes between a lipid-rich animal and its environment approach a steady state, the distribution of the contaminant in the tissues can be correlated with the lipid concentration within the tissues. Therefore, factors that affect lipid levels, such as seasonal storage cycles in digestive and reproductive tissues and the development of lipid-rich eggs, can affect the bioaccumulation and relative tissue distribution of contaminants (Gosling, 1992). In addition, spawning of lipid-rich eggs may represent a major route of loss of hydrophobic chemicals (Hummel et al., 1990).

Further, high POP concentrations found in mussels in different seasons by several authors (see above) have been mainly attributed to seasonal changes in the lipid content. Ferreira et al. (1990) found high concentrations of PCBs and DDTs in the oyster *Crassostrea angulata* in Portugal during winter and early spring, and they have attributed this to the high lipid content in those specimens during winter. Apart from total lipids, Ferreira and Vale (1998) found changes in the lipid composition in experimental clams and oysters. They observed a gradual decrease in reserve lipids in oysters and an alteration of lipid metabolism in smaller clams that had accumulated high concentrations of PCB residues during the experiment. The authors observed a pronounced consumption of triglycerides in PCB-exposed oysters; this confirmed the degradation of the physiological state of *Crassostrea angulata* exposed to high PCB concentrations during controlled experiments. Such changes were observed only in young clams having small body sizes. Clams exposed to high concentrations of PCB residues showed a reduction in triglycerides and an increase in sterol and fatty acids. After prolonged exposure of approximately 60 days, the authors observed a substantial decrease in the concentrations of PCB residues when the triglycerides were at low concentrations. According to the authors, these reductions may in turn be related to food ingestion rates. Tanabe et al. (1987) had previously established that PCBs are taken up by mussels mainly by equilibrium partitioning between PCBs in the ambient water and those in the lipid pool of the mussel. This process was noticed during both uptake and depuration processes. When the data were expressed on a lipid weight basis, very clear trends were observed in the data obtained during these processes.

The specimens of mussels collected from the coastal waters of China by Chen et al. (2002) had evidently higher OC concentrations in April than in October; the mussels were in a stage just prior to the reproductive phase. The animals with high lipid content at this stage of the life cycle naturally accumulated more lipophilic OCs in their bodies.

The effects of total lipids and lipid chemistry on the accumulation of hydrophobic chemicals such as POPs have been very clearly demonstrated ever since various animals have been used as bioindicators (Phillips, 1980), and this holds true till date (Gosling, 1996; Tanabe, 2000a; Ueno et al., 2003). Although the lipid content in mussels is not higher when compared with those in higher animals such as marine mammals, the normalization of the POP values on a lipid weight basis always nullifies the small differences that may arise due to seasonality, environmental parameters, etc. Further, the use of lipid weights as a basis for determining OC concentrations substantially decreases the variability between individuals in a given population, and this will permit a more sensitive definition of inter-sample differences encountered during monitoring surveys of these compounds (Phillips, 1980). As a result, it may be always advisable to express the data on POPs obtained from the organisms on both wet weight and lipid weight bases, and the measurement of extractable lipid concentrations in these organisms should be made mandatory. All the other parameters explained in the forthcoming sections may have some impact on the monitoring data on POPs, which can be rectified at different stages of monitoring (sampling, analysis, etc.). However, it should always be borne in mind that expression of data on a lipid weight basis will normalize many of the discrepancies that may arise during the regular course of monitoring.

Species Differences

The International Mussel Watch Project was undertaken under the auspices of the UNESCO and UNEP to assess the extent of chemical contamination in coastal areas, primarily in the equatorial and subequatorial regions of the

Southern Hemisphere, with particular attention to coastal areas in developing countries. In the initial implementation phase, an attempt was made to collect coexisting species of mussels and oysters in the same areas in South American countries. The concentrations of OCs such as DDTs, HCHs, CHLs and PCBs were compared in several species belonging to the genera *Crassostrea, Mytilus, Anadara, Mytella*, etc. (Farrington and Trip, 1995). Despite being exposed to the same concentrations of HCHs, CHLs, DDTs and PCBs within a habitat, the specimens of bivalves belonging to different species showed several small differences when the values of OC concentrations from the same sites or nearby sites were compared. Most tissue concentrations were found to vary by a factor of three or less. Earlier, O'Connor (1992) also found differences in the concentrations of total polyaromatic hydrocarbons (PAHs), DDTs, PCBs and CHLs in the mussels collected from the coastal areas of USA; these concentrations varied by a factor of two to three. In their report on the US National Status and Trends (NS&T) Mussel Watch Projects, Sericano et al. (1995) stated that the chemical analysis of coexisting species of bivalves at the same locations (i.e. exposed to the same contaminant loads) indicated that the concentrations of these contaminants in their tissues varied by a factor of four or less. These differences are small to permit a global comparison of POPs data even if different species of mussels are collected and analyzed.

In a transplantation experiment on the uptake kinetics of persistent OCs using blue and green mussels in Aburatsubo Bay and Tokyo Bay, Ueno et al. (1999) found that the concentrations of PCBs and DDTs in transplanted mussels reached those in native mussels and oysters within two weeks. The compositions of DDT- and CHL-related compounds in the transplanted mussels also changed to those in the native organisms. This suggested that these mussels had the ability to respond to changes in the ambient levels of OCs; hence, they were suitable as bioindicators for understanding short-term changes in pollution by POPs. As an important part of this study, the authors found that no significant difference was observed in the OC concentrations and compositions between the blue and green mussels in the overall transplantation experiment (Figs. 2.1 and 2.2). This implies that the use of a single species of mussels as an indicator is not a prerequisite in marine pollution monitoring.

Natural Factors
A central assumption in using bioindicators for monitoring temporal and spatial trends in POPs is that chemical concentrations in an organism depend linearly on those in the diet and in ambient water. This may not be the case in all the organisms. In the case of mollusks, evidences from laboratory (Pruell et al., 1987) and field experiments (Sericano, 1993) proved the existence of such

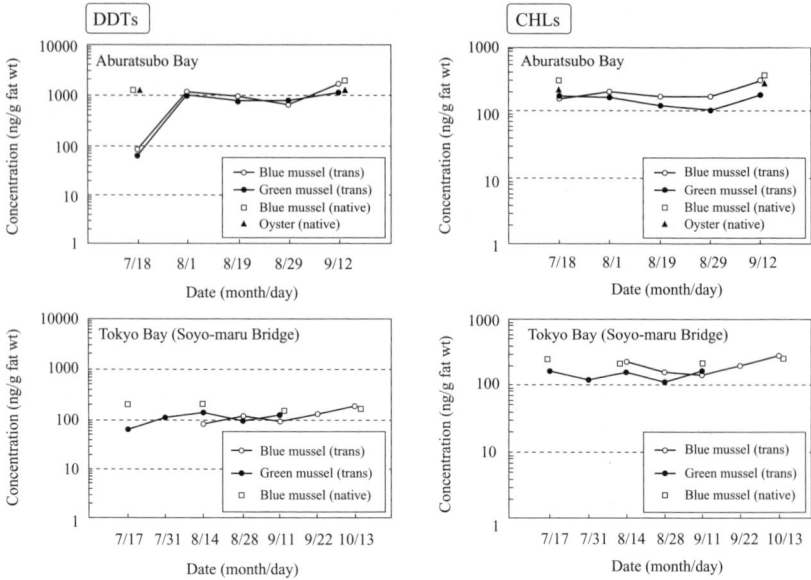

Fig. 2.1. Variations in the concentrations of DDT and CHLs in blue and green mussels during the transplantation experiments in the Aburatsubo Bay and Soyo-maru Bridge in Tokyo Bay. The concentrations of these compounds in native blue mussels and oysters are also shown for comparison with those in the transplanted mussels (Source: Ueno et al., 1999). (The term Trans in the figure indicates data from transplanted mussels).

linear relationships. However, these relationships are susceptible to various natural factors such as size of organism, location relative to tide, salinity, temperature and reproductive factors that can affect the concentrations of chemicals in mollusks (Phillips, 1980; Phillips and Segar, 1986). O'Connor (1996) found that it is difficult to nullify the effects of these natural factors on contaminant concentrations; therefore, they may affect the temporal and spatial trends. Collecting mollusks of a predetermined size at the same time of a year will diminish some natural causes of annual variation. If the sampling is performed by nullifying all the natural variations, which is possible in the case of mussels, the remaining trends in the data set can be easily interpreted based on anthropogenic contamination in the sampling area.

Seasonality

While analyzing the data set obtained from blue mussels (*Mytilus edulis* L.) from six sampling sites in the coastal regions of Germany, Huhnerfuss et al. (1995) noticed clear seasonal variations in the concentrations of PCB congeners 88, 149,

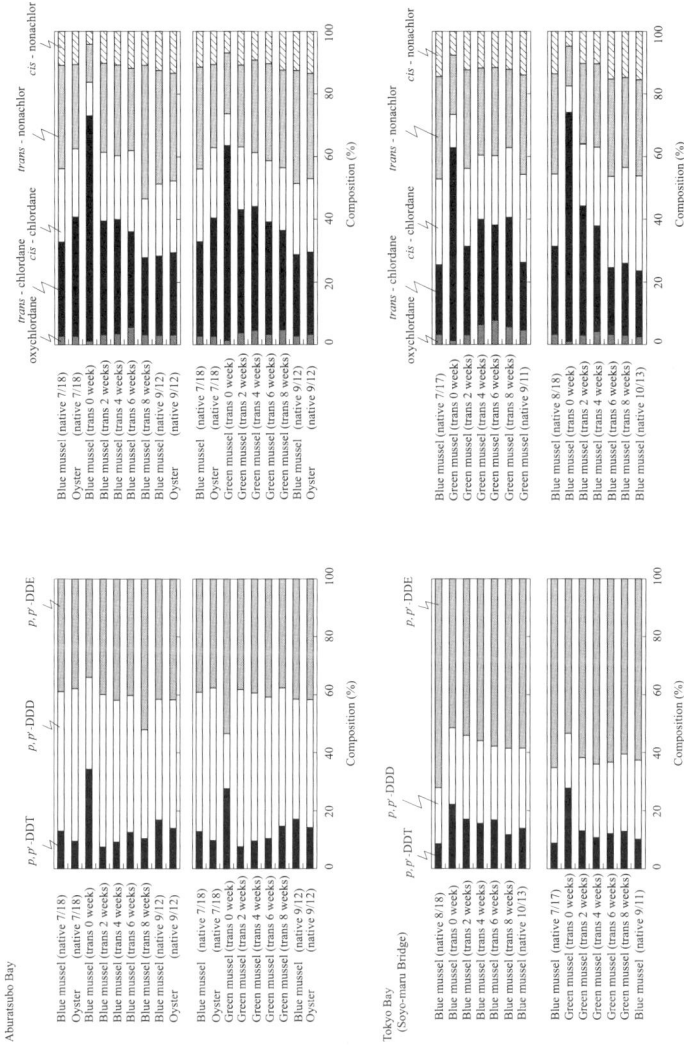

Fig. 2.2. *Variations in the composition of DDT and CHLs in blue and green mussels during the transplantation experiments in the Aburatsubo Bay and Soyo-maru Bridge in Tokyo Bay. The compositions of these compounds in native blue mussels and oysters are also shown for comparison with those in the transplanted mussels (Source: Ueno et al., 1999). (The term Trans in the figure indicates data from transplanted mussels).*

183, 174 and 171 between the specimens collected during autumn and spring. They found a maximum TEQ value of 1.5 ng/g on an extractable organic matter (EOM) basis and argued that such changes in mussels may be expected due to factors such as temperature changes, cycles in industrial activity and annual alterations in climate. In addition, since marine organisms themselves are highly seasonal in terms of their basic physiology and detoxification biochemistry, the concentrations of xenobiotics detected in tissue samples may follow a variable seasonal pattern (Sheeham and Power, 1999). The same authors have shown that the benzo(a)pyrene hydroxylase and NADPH-independent 7-ethoxycoumarin-O-deethylase (ECOD) activities in *Mytilus edulis* were higher in autumn than in spring.

Lee et al. (1996) observed a strong accumulation of HCHs, HCB, DDTs and PCBs in *Mytilus edulis* during winter. They observed strongly elevated levels of these compounds during winter, particularly in highly contaminated sites. On the other hand, this trend was less pronounced in less contaminated sites. It is a well-known fact that the lipid content in organisms is influenced by their physiological condition and reproductive cycle. The authors have also assumed that the seasonal fluctuations in organic contaminants in mussel tissues may be a result of different endogenous and exogenous factors such as availability of food, possible changes in the filtration rate during different seasons and seasonal changes in contaminant inflow. Ferreira et al. (1990) also found higher concentrations of PCBs and DDT in the tissues of the oyster *Crassostrea angulata* during winter/early spring than during other seasons. The authors suggested that biological changes such as the changes in lipid content as well as environmental variations in the contaminant levels could explain this temporal pattern. Both the above authors have reported data on the OC concentrations in mussels on a dry weight basis and have attributed such changes mainly to the reproductive cycles of these organisms rather than seasonality. Boon and Duinker (1986) and Ueno et al. (2001) stated that the OC concentrations in mussels when expressed on a lipid weight basis reflect changes in the ambient environment rather than seasonal physiological changes in the mussels. Therefore, the seasonal variations in OC concentrations in mussels appear to reflect the levels in the ambient water rather than seasonal physiological changes in the mussels, thus underlining the suitability of mussels as bioindicators for monitoring POPs in coastal ecosystems.

For most contaminants, it is apparent that bioaccumulation in mussels adequately reflects the changing levels in the environment (Phillips, 1980; Farrington et al., 1987; Cossa, 1988; Fowler, 1990). Typically, both abiotic and biotic factors can modify bioaccumulation by a factor of two. This magnitude of variation is negligible when detecting and comparing the marked spatial and temporal differences in contaminant concentrations in mussels. For discrimination with greater precision, the influence of factors that affect bioaccumulation may be considered while designing sampling strategies and interpreting results.

Age and Body Size

Bivalves have a reasonably long lifespan; therefore, specimens of different age classes (size ranges) can be sampled to study age-related accumulations. At the same time, care should be taken to include this parameter in the sampling schedule for spatial and temporal monitoring studies using bivalves. For example, Ferreira and Vale (1998) noticed considerable variations in PCB concentrations in clams of various size ranges, but not in oysters. In a 60-day laboratory experiment, the authors found that the accumulation of PCBs in smaller specimens (<25 mm) of clams of the species *Ruditapes decussatus* was 10 times higher than in larger specimens (>35 mm). However, the authors did not notice a considerable difference in the accumulation of PCBs between smaller (<50 mm) and larger (>60 mm) specimens of oysters of the species *Crassostrea angulata*. Apart from reporting data on a dry weight basis, the authors have suggested various reasons such as higher feeding rate in smaller animals and alterations in assimilation rates for the observed variation in PCB concentrations. The authors have also suggested that this relationship between OC accumulation and size may not hold true in all species of mussels. For example, in the oyster *Crassostrea angulata*, they found uniformity in lipid content and size independence with respect to assimilation rate as well as comparable PCB concentrations between two body size ranges. In yet another species, *Crassostrea gigas*, Gerdes (1983) found size-independent assimilation efficiency.

Further, while examining the mussel *Mytilus galloprovincialis* collected from Tokyo Bay for spatial variations in OCs such as PCBs, DDTs, CHLs, HCHs and HCB, Ueno et al. (2001) found that neither shell size nor vertical habitation of mussels appeared to affect the concentrations expressed on a lipid weight basis. They noted that the seasonal variations in OC concentrations on a lipid weight basis appeared to reflect the levels in ambient water rather than seasonal physiological changes in mussels. These results suggest that the variable environmental or physiological factors did not significantly affect the availability and suitability of mussels as bioindicators for monitoring of contamination by POPs. They also reported that differences observed in the concentrations of PCBs, DDTs and CHLs in the tissues of mussels with larger shells (>45mm) and those with smaller shells (<45mm) during different months of sampling were not significant. This proves that rather than the actual size of the organisms, the POP concentrations in mussels reflect the levels in the ambient environment (Fig. 2.3).

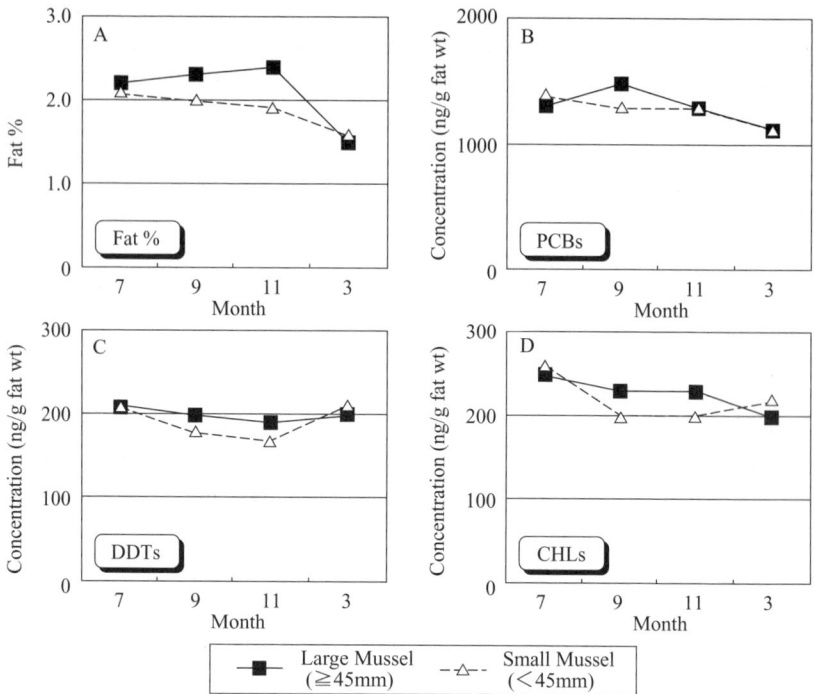

Fig. 2.3. Seasonal variations in OC concentrations and lipid contents in blue mussels collected from the mid-tidal level (Source: Ueno et al., 2001).

Further, as stated by Gosling (1992), small variations that may occur due to the body size of the organisms can be minimized by sampling 'standard body size' mussels.

Uptake from Different Fractions

The concentrations of lipophilic pollutants in gill-breathing aquatic animals depend primarily on their levels in water. Uptake of additional amounts of pollutants through ingestion of contaminated food does not greatly influence the total concentrations attained in an organism. Further, experiments with *Mytilus edulis* revealed that the concentrations of pollutants in these organisms showed excellent agreement with the bioconcentration factors (BCFs) for PCBs and PAHs calculated using the dissolved phase concentrations; however, this was not observed for the BCFs calculated using the concentrations in the contaminated sediment (Pruell et al., 1986). Even with crude oil contamination, clams suspended above the bedded sediment showed greater accumulation of these contaminants than those placed on the sediment. This shows that the dissolved phase above the sediment is the source of contaminants for clams.

Ueno et al. (2001) found that in *Mytilus galloprovincialis*, the vertical habitation of the organism did not affect the accumulation patterns of PCBs, DDTs and CHLs. They collected specimens from both high and low tide areas of Tokyo Bay and did not find any notable difference in the POP concentrations. This indicates that bivalves inhabiting different depths can be utilized as indicators for measuring temporal and spatial variations in POPs.

Metabolism

It has been reported that cytochrome P450 (CYP) enzyme activities are not inducible in the bivalve digestive gland, which is in contrast to that observed in the liver of vertebrate fish where the enzyme activities are readily induced by xenobiotics (Livingstone, 1991; Livingstone et al., 1995). As a result, the levels of CYP activity are generally lower in *Mytilus edulis* than in mammalian species (Sheehan and Power, 1999). Livingstone et al. (1995) observed site differences in certain enzyme activities in the liver of goby with changes in tissue contaminant (PAHs, PCBs and DDTs) concentrations. On the other hand, mussels showed little site difference in NADPH-cytochrome c reductase and P-450 activities.

The rates of uptake and final tissue concentrations of organic xenobiotics are mainly determined by the exposure concentrations in the ambient water and the metabolic capacity of the organism. Livingstone et al. (1990, 1992) found that the rates of metabolism for a given concentration of xenobiotics in particular tissues are lower in mollusks than in crustaceans. Using pooled literature data, the authors showed that in mollusks, the uptake of hydrocarbons generally exceeds the rates of metabolism by an order of magnitude or more; this difference accounts for the marked bioaccumulation of these compounds. This process can be explained using the simple lipid/water equilibrium model (Burns and Smith, 1981; Tanabe et al., 1987). The fundamental reason for this phenomenon is presumed to be as follows. Although uptake is mainly passive and determined by physicochemical principles, metabolism (e.g. enzyme activities) is intrinsic and mainly determined (limited) by endogenous factors, i.e. energy costs and endogenous functions of biotransformation enzymes (Gosling, 1992). Other factors that undoubtedly contribute to the long residence times of organic xenobiotics in mollusks are a slow release of metabolites into seawater and the covalent binding of xenobiotics to macromolecules (Livingstone, 1990, 1992). Thus, a low metabolic capacity, slow action of metabolic enzymes and slow release of metabolites into ambient water renders mollusks, which are a step superior to other invertebrates, apt bioindicators for monitoring POPs in coastal environments.

Monitoring Studies

Because of the ease with which mollusks can be used as biomonitors of coastal pollution and the rapid growth and worldwide application of Mussel Watch Programs, it may not be possible to provide a thorough review of all the POP monitoring programs that use mollusks. Therefore, it may be appropriate to review some of the comprehensive and most recent studies carried out in different regions of the world by using various bivalve mollusks. Reviews on Mussel Watch Programs cited at the end of this chapter will help the readers to obtain more information on this issue.

The US National Oceanic and Atmospheric Administration (NOAA) initiated the NS&T Program in 1984 in order to determine the current status and temporal trends in the environmental quality of the US coastal and estuarine waters. A major part of this program involves monitoring the concentrations of trace and major elements and organic contaminants in benthic fish, bivalve mollusks and sediments (NOAA, 1993). The NS&T Program's Mussel Watch Project was initiated in 1986 and at present monitors a suite of contaminants in the tissues of bivalve mollusks (mussels and oysters) and in sediments at 245 sites along the coastal and estuarine waters of the US. As a part of this project, molluskan tissue samples are monitored annually for approximately 70 contaminants, including 24 PAHs; 20 congeners and isomers of PCBs; 15 chlorinated pesticides, including CHLs and DDTs; butyltins; 4 major elements and 12 trace elements. The California Department of Fish and Game's State Mussel Watch Program (SMWP) has been in effect since 1976. It is a part of a worldwide monitoring effort aimed at detecting the concentrations of trace elements, pesticides and PCBs by analyzing native and transplanted mussels and clams in Californian waters. The information gathered by these programs was used to track the temporal changes and geographic distribution of toxic substances and to identify potential problem areas where more intensive studies are necessary.

Using the annual monitoring data on the mussel specimens collected along the US coast during 1986–1993 under the NS&T Mussel Watch Project, O'Connor (1996, 1998) and Beliaeff et al. (1997) found that the levels of chlorinated hydrocarbons decreased along the US coasts following the ban on their usage. They have stated that these trends were not unexpected because the use of all the monitored chlorinated hydrocarbons had been banned in the US since many years prior to the sampling. For example, in the US, PCBs began to be phased out in 1971, and DDT was banned in 1972; dieldrin, in 1975 and CHL,

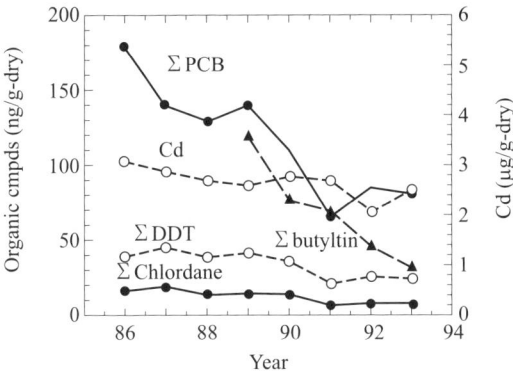

Fig. 2.4. Annual geometric concentrations of chemicals in mollusks from sites from 1986 to 1993 (Source: O'Connor, 1996).

in 1983. The effects of these steps are clearly reflected in the data collected from mussels (Fig. 2.4), indicating the suitability of mussels for measuring the annual trends in POPs.

Total concentrations of DDTs, CHLs, PCBs and PAHs showed a fairly homogenous distribution along the northern coast of the Gulf of Mexico. But the results of the NS&T and International Mussel Watch Programs (1986–1993) conducted along the North, Central and South American coasts showed a larger variability in the geographical distribution of some of these contaminants. For example, high concentrations of DDT and its metabolites DDD and DDE were found in the tropical and subtropical areas than in more temperate zones of South America (Sericano et al., 1995). Although different species of mussels were sampled under the International Mussel Watch Program (1991–1992) because of the large study area in the Americas, the authors could find clear spatial variations in the distributions of compounds such as PCBs and PAHs; the highest concentrations of these compounds were encountered along the southern South Atlantic coast.

The initial phase of the Mediterranean Mussel Watch Project (CIESM, 2002) documented spatial and temporal trends in heavy metals and radioactivity in the Mediterranean waters along the coastal areas of several countries such as Egypt, Syria, Spain, Morocco, Greece, Croatia, Algeria, Tunisia, France, Bulgaria and Turkey; the measurements were based on the total body burden of filter-feeding mollusks, preferably mussels of the order Mytilidae. The authors have also suggested that the ultimate aim of the next phase is to extend the program to other trace contaminants such as POPs in accordance with a common global approach of other mussel watch projects (CIESM, 2002). The results of the program emphasized the importance of mussels as bioindicators of trace contaminants in coastal waters.

Several species of bivalves were used as bioindicators of the spatial and temporal variations in POPs. Chen et al. (2002) tested the concentrations of HCH, DDTs and PCB residues in *Crassostrea gigas, Crassostrea plicatula, Ruditapes philippinarum, Mytilus viridis, Mytilus coruscus* and *Meretrix meretrix* collected from the areas near Xiamen Island and Minjiang Estuary, China. They found bioaccumulation of all the three compounds in all the analyzed animals. The concentrations in these animals were apparently one or two orders of magnitude higher than in the nearby sediments; further, the concentrations of these compounds in the sediments were linearly reflected in bivalves, showing that the concentrations in bivalves could reflect the status of environmental contamination. The oysters were found to be good bioaccumulators of OCs among the analyzed species; the concentrations of all the analyzed compounds were higher in oysters than in the other species. The authors also found higher concentrations of DDTs than HCHs and PCBs in mussels and oysters; this was also observed by Tanabe (2000a) and Monirith et al. (2003) in the green mussels (*Perna viridis*) collected from the coastal waters of Hong Kong and China. Of the six species analyzed, Chen et al. (2002) concluded that oysters may be a more ideal bioindicator for OC monitoring.

As a part of the Mussel Watch Program, Kim et al. (2002) collected specimens of the oyster *Crassostrea gigas*, mussels *Mytilus edulis* and *Mytilus coruscus* and clams *Cyclina sinensis* and *Ruditapes philippinarum* from 66 sites along the east, west and south coasts of South Korea in 1999 and analyzed them for PCBs, DDTs, CHLs and HCHs. PCBs were predominant along the Korean coast, followed by DDTs, HCHs and CHLs. The authors observed spatial variation in PCBs, DDTs and CHLs in the coastal areas, indicating the presence of terrestrial input pathways, whereas the even distribution of HCHs suggested an atmospheric pathway to coastal waters. Global comparison showed that the Korean coasts had lower levels of these compounds than the US coasts but somewhat higher levels than the coasts of Asian developing countries, except for some compounds having specific contamination sources in some countries (Tanabe et al., 2000).

The Asia-Pacific Mussel Watch Program (APMW) was conducted during 1997–2000; it monitored marine pollution in this region by using mussels as bioindicators (Tanabe, 1994, 2000b). This project was a part of the International Mussel Watch—Asia Pacific Phase. The project's principal activity was coastal monitoring by using sentinel bivalves such as mussels and oysters as bioindicators in order to ascertain the quality of coastal marine waters. The initial phase of the program was conducted in South and Central America during 1991–1993 and revealed serious contamination by OC insecticides in the third world countries in this region. The APMW was a collaborative work of scientists from Cambodia, China, Hong Kong, India, Indonesia, Japan, Korea, Malaysia, the Philippines, Far East Russia, Singapore and Vietnam; Prof. Tanabe of Ehime University, Japan, was the project leader. The project intended to cover the three zones of the Asia-Pacific region, namely, the northwest North Pacific, the tropical regions of the ASEAN countries and the South Pacific.

Monirith et al. (2003) summarized the contamination status, distribution and possible pollution sources of OCs such as PCBs, DDTs, CHLs and HCB as well as HCHs as an outcome of the APMW. They detected OCs in all the mussel samples from all the sites investigated in all the countries. Considerable concentrations of p,p'-DDT and α-HCH residues were found in many of the mussels, and the concentrations of DDTs and HCHs found in the mussels from Asian developing countries were higher than in those from developed nations; this suggests current usage of DDTs and HCHs along the coastal waters of Asian developing countries. On the other hand, detection of lower PCB concentrations in the mussels from Asian developing countries than in those from developed nations indicates that PCB contamination in mussels is strongly related to industrial activities.

Based on the residue patterns in *Perna viridis* from Asian countries, Monirith et al. (2003) found that the mussels from China, Hong Kong and Vietnam had higher concentrations of DDTs; India, China and Russia, higher concentrations of HCHs; Japan, Russia, Hong Kong and Malaysia, higher PCB concentrations and Japan, China, Hong Kong and Malaysia, higher concentrations of CHLs (Fig. 2.5). Interestingly, the authors found that the mean and 90th percentile values of the concentrations agree well with the per capita gross national product (GNP) of each country. Since the GNP value is an indicator of economic status and PCB and CHL contaminations are strongly related to industrial and human activities, concentrations of PCBs and CHLs may increase in countries with high economic growth rates. Based on their studies with mussels, the authors could chart the residue patterns of OC contamination in Asian countries. The authors suggested that these countries act as prime sources of pollution by these chemicals in this region.

The findings of APMW using *Perna viridis* as the bioindicator revealed widespread and country-specific OC pollution across the Asian developing countries. The authors of all the reviews based on the APMW project (Tanabe, 2000a; Sudaryanto et al., 2002; Monirith et al., 2003) suggested the necessity of performing continuous monitoring as well as further studies to elucidate the

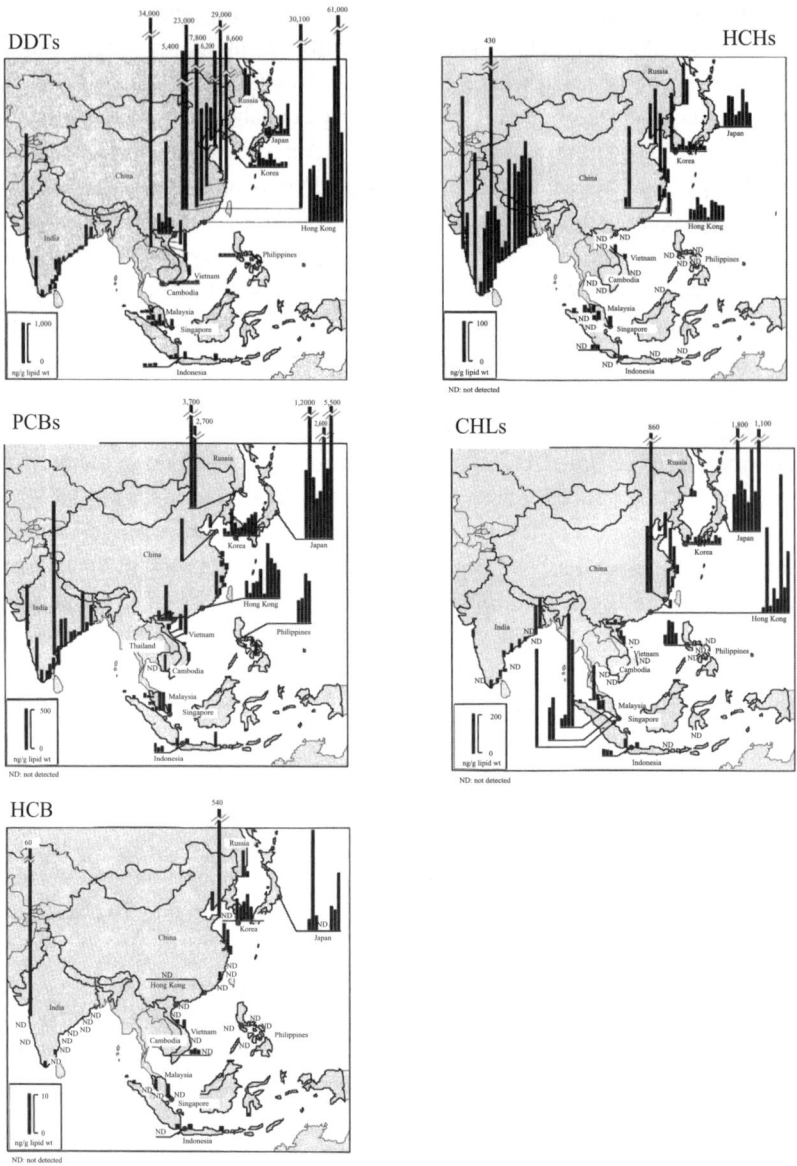

Fig. 2.5. Distribution of concentrations of DDTs, HCHs, PCBs, CHLs and HCB in mussels collected from the coastal waters of some Asian countries (Source: Monirith et al., 2003).

trends in contamination and its impact on organisms for ecotoxicological risk assessment. The authors also raised concerns regarding the lack of scientific information pertaining to the area and the urgent need for the capacity development of highly trained scientists in the countries within the Asia-Pacific region.

Mussels of the genus *Perna*, particularly the green mussel *Perna viridis*, are in many respects highly suitable for use as a bioindicator organism in the tropical and subtropical Asia-Pacific region, mainly because of its wide distribution in that area. Further, this species is recognized as a commercially valuable seafood. Besides oyster culture, this species is one of the important species of bivalves currently being commercially cultured outside Europe. In a recent study on the tolerance of five species of tropical marine mollusks (*Perna viridis, Perna perna, Barchidontes straitulus, Brachidontes variabilis* and *Modilus philippinarum*) to continuous chlorination, *Perna viridis* was found to be the most tolerant species that withstood 1 mg/l of continuous chlorination for 816 h. On the other hand, the least tolerant species *Brachidontes variabilis* could tolerate continuous chlorination for only 288 h. *Perna viridis* was the most tolerant species at all chlorine levels tested (Rajagopal et al., 2003). Among the five species, the order of tolerance to chlorination was *Perna viridis* > *Brachidontes striatulus* > *Perna perna* > *Modilus philippinarum* > *Brachidontes variabilis*. It was also found that larger mussels were more tolerant than smaller individuals. This study underscores the suitability of the species *Perna viridis* as a bioindicator in natural environments under conditions of stress such as high contamination levels of OC pesticides.

There have been innumerable attempts by using different bivalves as bioindicators of POP pollution; bivalves belonging to the genera *Mytilus, Crassostrea* and *Perna* are most commonly used as bioindicators. Recently, there is a growing trend toward using the species belonging to the genus *Perna*, particularly *Perna viridis*, popularly known as green mussels or green-lipped mussels, inhabiting tropical and subtropical waters. Reviewing some of the recent works on the usage of *Perna viridis* at this juncture may be appropriate and useful because this book aims at recommending suitable bioindicators for monitoring of POPs in the tropical and subtropical regions where most of the developing countries are situated.

Transplantation Experiments

Apart from field studies, the use of *in situ* bioassays using caged bivalves is a potentially powerful tool in the risk assessment process for making these direct measurements (Salzar and Salzar, 1998). Caging bivalves facilitates the quantification of exposure and response under natural conditions. This reduces the uncertainty associated with extrapolating the results of laboratory bioassays conducted under artificial exposure conditions.

Bivalves have the ability to accumulate contaminants from ambient water and to concentrate these to higher levels; hence, using them in such *in situ* experiments may provide a direct link between exposure, dose and response.

Many species of freshwater and marine bivalves have been used successfully in *in situ* field bioassays, but the majority of the work has focused on mussels because of their availability, distribution and sensitivity. Transplanted populations of bivalves have been used as biomonitors of environmental contamination for the past 30 years, and the use of marine bivalves such as *Mytilus* sp. has been more extensive (MacMahon, 1991). The US Fish and Wildlife Service has identified several mussel, oyster, clam and scallop species as possible surrogate species for *in situ* field testing in lieu of threatened and endangered species (Koch, 1999).

Salzar and Salzar (1996a, 1996b) and Bridgman (1999) have published several reports on the Caged Mussel Pilot Study in which different species of mussels were successfully used in *in situ* experiments for monitoring the environmental effects of pulp and paper mills in Canada. These experiments measured the concentrations of certain organic pollutants such as tributyltin (TBT) and PAHs in Alaska Bay. Detailed information regarding the design and conduct of the *in situ* experiments by using caged bivalve mollusks is available on the websites of the authors. Based on their extensive experiments, the authors have stated that the results of *in situ* field bioassays may be more representative of environmental effects than those of laboratory bioassays because the animals are exposed to all site-specific stressors. Estimating bioaccumulation by using caged animals may thus minimize the assumptions that are necessary with conventional field and laboratory approaches and provides information that is more useful.

Caging of bivalves facilitates growth measurement that enables the estimation of growth-related bioaccumulation. Because bioaccumulation measurements are performed synoptically, the data provide a direct link between exposure and dose. These manipulative field studies are useful in bioindicator studies because they bridge the gap between standard laboratory bioassays and traditional monitoring of resident populations. *In situ* bioassays include the advantages of both the approaches. Moreover, *in situ* field bioassays using caged bivalves may allow monitoring individual organisms as well as sampling an almost infinite matrix of space and time because the animals can be strategically placed along physical and chemical gradients associated with both the water column and sediments. Utilizing animals with a known history of biological parameters and chemical exposure may facilitate easy and accurate data interpretation. Wherever necessary (i.e. when adequate numbers of specimens are not available), caged bivalve studies can be conducted by transplanting animals, which are suitable for assessing the exposure, from nearby pristine areas.

Conclusions

Bivalve mollusks have been extensively used to assess the levels of contamination by numerous environmental contaminants, including POPs. In marine ecosystems, certain genera and species, notably those of mussels and oysters, have been extensively studied, particularly in temperate waters. Bioaccumulation by these organisms provides a time-integrated measure of contaminant bioavailability, responding essentially to the fraction of total environmental load that is of direct ecotoxicological relevance. Bivalves are definitely superior to other marine organisms for use as bioindicators of POP monitoring. As exemplified by their suitable biological and ecological characteristics, many of the bivalve species, particularly mussels and oysters, appear to be highly suitable for use as bioindicators in tropical developing countries. The species belonging to the genus *Perna*, particularly the green mussel *Perna viridis*, appear to be promising candidates for POP monitoring in the coastal waters of developing countries due to their predominant distribution in the tropical and sub-tropical regions where most of the developing countries are situated. In the coastal and estuarine areas of Asian developing countries, candidate species belonging to the genus *Perna* are widely distributed mostly in large populations throughout the year. As explained earlier in this chapter, mussels belonging to the genus *Mytilus* and oysters of the genus *Crassostrea* also serve as good bioaccumulators of POPs, and there are only marginal differences in the bioaccumulation between these genera. Therefore, in case of non-availability of *Perna viridis*, other species of the genus *Perna*, and in their absence, species of the genera *Mytilus* and *Crassostrea,* in that order of preference, can be gathered and evaluated. Simultaneously, attempts should be made to limit the number of species that are collected to enable reasonable comparison of chemical concentration data over the largest possible geographic area. The bioaccumulation characteristics may differ depending on age, sex, reproductive state, feeding, field conditions, etc.; most of the differences can be rectified by developing an appropriate sampling protocol.

The NOAA (1993) has formulated several criteria for the selection of mussel watch sites and species, and it would be appropriate to cite some of these criteria at this juncture. Since the purpose of the present discussion is to recommend suitable bioindicators for regular monitoring, collecting mollusks from sites that are representative of natural surroundings would be better than collecting them from known 'hot-spots' of POPs. Known or suspected hot spots of POPs may be avoided, unless their monitoring is required for specific purposes.

A literature survey may be conducted to select and recommend a suitable bivalve species for use as a bioindicator of POPs in all the regions included in a particular program. Either a single species or two to three related species may be recommended. Historically, oysters and mussels have served as good indicator organisms. These organisms may be selected preferably in places where their native populations are abundantly available throughout the year.

A sufficient number of specimens should be available so that a level of statistical significance can be assigned to the data. At least 20–30 surviving specimens should be available at each sampling time to permit pooling of tissues, if required. Native populations of mollusks must inhabit the sampling site because caged and transplanted mussels are not used in regular monitoring surveys. The sampling site should be free of human interference (e.g. dredging and recreation) to avoid disturbance of natural processes. Both water and sediment samples should preferably be collected from monitoring areas during each sampling event and analyzed for the target pollutant. All water quality parameters such as temperature, salinity and pH should be monitored. Several fluctuations may occur in the POP uptake by bivalves. Care should be taken to monitor the 'peak' and 'lean' accumulation months. These fluctuations may occur due to variations in seasonal (winter and summer) or biological (reproduction and feeding) parameters; all such parameters should be considered while planning sampling strategies at individual sites. If native populations do not qualify as good bioindicators, the option of using representative species from nearby, identical but clean environment may be considered.

For capacity development, training programs involving the use of bivalves as biomonitors and transplantation experiments can be conducted for scientists of all developing countries. The scientists may be given the liberty to handle any situation in the field and to select specimens (within the prescribed criteria). All the field staff and scientists should be trained and updated (based on field experiences) on the complex, multiphase and multimatrix approaches of using bivalves as POP biomonitors.

Further, the currently available knowledge, which was obtained through various Mussel Watch Projects conducted in several regions of the world, may be extremely useful in planning and executing field and transplant studies by using bivalves as bioindicators of POPs. Reviews on mussel watch and transplantation experiments by several authors such as Martin (1985), Farrington and Tripp (1995), O'Connor (1996, 1998), Salzar and Salzar (1996a, b), Bridgman (1999) and Tanabe (2000a) can demonstrate the feasibility and scientific value of using mussels as a monitoring species for evaluating POP exposure and the associated environmental effects.

References

Beliaeff, B., T.P. O'Connor, D.K. Daskalakis and P.J. Smith, 1997. Mussel Watch data from 1986 to 1994: Temporal trend detection at large spatial scales. Environ. Sci. Technol., 31: 1411–1415.

Boon, J.P. and J.C. Duinker, 1986. Monitoring of cyclic organochlorines in the marine environment. Environ. Monit. Assess., 7: 189–208.

Bridgman, J., 1999. Caged Mussel Monitoring Pilot Study. Ballast Water Treatment Facility, Final Report, Project No. RCAC – PV1, p. 87.

Buck, W.B., 1979. Animals as monitors of environmental quality. Vet. Hum. Toxicol., 21: 277–284.

Burns, K.A. and J.L. Smith, 1981. Biological monitoring of ambient water quality: the case for using bivalves as sentinel organisms for monitoring petroleum pollution in coastal waters. Estuar. Coast. Shelf Sci., 13: 433–443.

Butler, P.A., 1966. Pesticides in the marine environment. J. Appl. Ecol., 3: 253–259.

Cain, B.W., 1981. Nationwide residues of organochlorine compounds in wings of adult mallards and black ducks, 1979–80. Pestic. Monit. J., 15: 128–134.

Chen, W., L. Zhang, L. Xu, X. Wang, L. Hong and H. Hong, 2002. Residue levels of HCHs, DDTs and PCBs in shellfish from coastal areas of east Xiamen and Minjiang estuary, China. Mar. Pollut. Bull., 45: 385–390.

CIESM, 2002. Mediterranean Mussel Watch-Designing a regional program for detecting radionuclides and trace contaminants. CISEM Workshop Series, No. 15, p. 136, Monaco. <www.cisem.org/publications/Marseilles02.pdf>

Cossa, D., 1988. Cadmium in *Mytilus* spp.: Worldwide survey and relationship between seawater and mussel content. Mar. Environ. Res., 26: 265–284.

Dewitt, J.B., C.M. Menzie, V.A. Adomaitis and W.L. Reichel, 1960. Pesticidal residues in animal tissues. Trans. N. Am. Wildl. Nat. Resour. Conf., 25: 277–285.

Farrington, J.W. and B.W. Trip, 1995. NOAA Technical memorandum NOS ORCA 95. International Mussel Watch Project. Initial Implementation Phase, Final Report, US Dept. Commerce, NOAA National Oceanic and Atmospheric Administration, p. 63.

Farrington, J.W., A.C. Davis, B.W. Tripp, D.K. Phelps and W.B. Galloway, 1987. 'Mussel Watch' – Measurement of chemical pollutants in bivalves as one indicator of coastal environmental quality. In: T.P. Boyle (Ed.), New Approaches to Monitoring Aquatic Ecosystems, ASTM STP 940, American Society for Testing and Materials, Philadelphia, pp. 125–139.

Ferreira, A.M. and C. Vale, 1998. PCB accumulation and alterations of lipid in two length classes of the oyster *Crassostrea angulata* and of the clam *Ruditapes decussatus*. Mar. Environ. Res., 45: 259–268.

Ferreira, A.M., C. Cortesa, O.G. Castro and C. Vale, 1990. Accumulation of metals and organochlorines in tissues of the oyster *Crassostrea angulata* from the Sado estuary, Portugal. Sci. Total Environ., 97: 627–639.

Fowler, S.W., 1990. Critical review of selected heavy metal and organochlorine hydrocarbon concentrations in the marine environment. Mar. Environ. Res., 29: 1–64.

Gerdes, D., 1983. The Pacific oyster *Crassostrea gigas*. Part I. Feeding behaviour of larvae and adults. Aquaculture, 31: 195–219.

Goldberg, E.D., 1975. The Mussel Watch: A first step in global marine monitoring. Mar. Pollut. Bull., 6: 111.

Gosling, E., 1992. The Mussel *Mytilus*: Ecology, Physiology, Genetics and Culture. Elsevier, Amsterdam, p. 589.

Huhnerfuss, H., B. Pfaffenberger, B. Gehrcke, L. Karbe, W.A. Konig and L. Landgraff, 1995. Stereochemical effects of PCBs in the marine environment: Seasonal variation of coplanar and atropisomeric PCBs in blue mussels (*Mytilus edulis* L) of German Bight. Mar. Pollut. Bull., 30: 332–340.

Hummel, H., R.H. Bogaards, J. Nieuwenhuize, De Wolf and L.J.M. Van, 1990. Spatial and seasonal differences in the PCB content of mussel *Mytilus edulis*. Sci. Total Environ., 92: 155–163.

Johnson, R.E., T.C. Carver and E.H. Dustman, 1967. Residues in fish, wildlife and estuaries. Pestic. Monit. J., 1: 7–13.

Kim, S.K., J.R. Oh, W.J. Shim, D.H. Lee, U.H. Kim, S.H. Hong, Y.B. Shin and D.S. Lee, 2002. Geographical distribution and accumulation of organochlorine residues in bivalves from coastal areas of South Korea. Mar. Pollut. Bull., 45: 268–279.

Koch, L., 1999. Surrogates for threatened and endangered species toxicity testing. US Fish and Wildlife Service, Southwest Field Office, Abingdon, Virginia.

Lee, K.M., H. Kruse and O. Wassermann, 1996. Seasonal fluctuation of organochlorine in *Mytilus edulis* L. from the South West Baltic Sea. Chemosphere, 32: 1883–1895.

Livingstone, D.R., 1991. Organic xenobiotic metabolism in marine invertebrates. In: R. Gilles (Ed.), Adv. Comp. Environ. Physiol., 7: 45–185.

Livingstone, D.R., 1992. Persistent pollutants in marine invertebrates. In: C.
Walker and D.R. Livingstone (Eds.), Persistent Pollutants in Marine
Ecosystems, Pergamon Press, Oxford.

Livingstone, D.R., G.P. Martinez, S. O'Hara, X. Michel, J.F. Nar Bonne, D.
Ribera and G.W. Winston, 1990. Oxyradical production as a pollution-
mediated mechanism of toxicity in the common mussel *Mytilus edulis* L.,
and other mollusks. Funct. Ecol., 4: 415–424.

Livingstone, D.R., F. Lips, P.G. Martinez and R.K. Pipe, 1992. Antioxidant
enzymes in digestive gland of the mussel *Mytilus edulis*. Mar. Biol., 112:
265–276.

Livingstone, D.R., P. Lemaire, A. Matthews, L.D. Peters, C. Porte, P.F.
Fitzpatrick, L. Forlin, C. Nasci, V. Fossato, N. Wootton and P. Goldfarb,
1995. Assessment of the impact of organic pollutants on goby (*Zosteris-
esor ophiocephalus*) and mussel (*Mytilus galloprovincialis*) from the
Venice Lagoon, Italy: Biochemical studies. [Mar. Environ. Res., 39:
235–240.

Martin, M., 1985. State Mussel Watch: Toxic Surveillance in California. Mar.
Pollut. Bull., 4: 140–146.

McMahon, R.F., 1991. Ecology of an invasive pest bivalve, *Corbicula*. In: The
molluska, Vol. 6, Ecology, Academic Press, pp. 505–561.

Monirith, I., D. Ueno, S. Takahashi, H. Nakata, A. Sudaryanto, A.N.
Subramanian, S. Karuppiah, A. Ismail, M. Muchtar, J. Zheng, B.J.
Richardson, M. Prudente, N.D. Hue, T.S. Tana, A.V. Tkalin, S. Tanabe,
2003. Asia-Pacific mussel watch: monitoring contamination of persistent
organochlorine compounds in coastal waters of Asian countries. Mar.
Pollut. Bull., 46: 281–300.

Moore, N.W., 1966. A pesticide monitoring system with special reference to the
selection of indicator species. J. Appl. Ecol., 3 (Suppl.): 261–269.

NOAA, 1993. National Status and Trends Program: Monitoring site descriptions
(1984–1990) for the national mussel watch and benthic surveillance
projects. NOAA Technical Memorandum NOS ORCA 70, p. 357.

O'Connor, T.P., 1992. Mussel Watch. Recent Trends in Coastal Environmental
Quality. NOAA, Rockville, Maryland.

O'Connor, T.P., 1996. Trends in chemical concentrations in mussels and oysters
collected along the US coast from 1986 to 1993. Mar. Environ. Res., 41:
182–200.

O'Connor, T.P., 1998. Mussel watch results from 1986 to 1996. Mar. Pollut.
Bull., 37: 14–19.

Phillips, D.J.H., 1980. Quantitative Aquatic Biological Indicators. Their Use to Monitor Trace Metal and Organochlorine Pollution, Applied Science Publishers, London, p. 488.

Phillips, D.J.H. and D.A. Segar, 1986. Use of bio-indicators in monitoring conservative contaminants: program design imperatives. Mar. Pollut. Bull., 17: 10–17.

Pruell, R.J., J.L. Lake, W.R. Davis and J.G. Quinn, 1986. Uptake and depuration of organic contaminants by blue mussels (*Mytilus edulis*) exposed to environmentally contaminated sediment. Mar. Biol., 91: 497–507.

Pruell, R.J., J.G. Quinn, J.L. Lake and W.R. Davis, 1987. Availability of PCBs and PAHs to *Mytilus edulis* from artificially resuspended sediments. In: J.M. Capuzzo and D.R. Kester (Eds.), Oceanic Processes in Marine Pollution, Vol. 1, Biological Processes and Wastes in the Ocean, Krieger, Malabar, Florida, pp. 97–108.

Rajagopal, S., V.P. Venugopalan, C. van der Velde and H.A. Jenner, 2003. Tolerance of five species of tropical marine mussels to continuous chlorination. Mar. Environ. Res., 55: 277–291.

Salzar, M.H. and S.M. Salzar, 1996a. Using caged bivalves for environmental effects monitoring at pulp and paper mills: Rationale and historical perspective. In: J.S. Goudey, M.D. Tressmann and A.J. Nimmi (Eds.), Proc. 23rd Aquatic Toxicology Workshop: Tools for Ecological Risk Assessment, Oct. 7–9, 1996, Calagary, Alberta, pp. 129–136.

Salzar, M.H. and S.M. Salzar, 1996b. Mussels as Bioindicators: Effects of TBT on Survival, Bioaccumulation and Growth Under Natural Conditions. In: M.A. Champ and P.F. Seligman (Eds.), Organotin, Chapmann & Hall, London, pp. 305–330.

Salzar, M.H. and S.M. Salzar, 1998. Using caged bivalves as part of an exposure-dose-response triad to support an integrated risk assessment strategy. In: A. de Peyster and K. Day (Eds.), Proc. Ecological Risk Assessment: A Meeting of Policy and Science, SETAC Press, Pensacola, Florida, pp. 167–192.

Sericano, J.L., 1993. The American Oyster (*Crassostrea virginica*) as a Bioindicator of Trace Organic Contamination. Ph.D. Thesis. Texas A & M University, College Station, Texas, p. 242.

Sericano, J.L., T.L. Wade, T.J. Jackson, J.M. Brooks, B.W. Tripp, J.F. Farrington, L.D. Mee, J.W. Readmann, J.P. Villeneuve and E.D. Goldberg, 1995. Trace organic contamination in the Americas: An overview of the US National Status & Trends and the International 'Mussel Watch' programs. Mar. Pollut. Bull., 31: 214–225.

Sheehan, D. and A. Power, 1999. Effects of seasonality on xenobiotic and antioxidant defence mechanisms of bivalve mollusks. Comp. Biochem. Physiol., C: Toxicol. Pharmacol., 123: 193–199.

Sudaryanto, A., S. Takahashi, S.I. Monirith, A. Ismail, M. Muchtar, J. Zheng, B.J. Richardson, A.N. Subramanian, M. Prudente, N.D. Hue and S. Tanabe, 2002. Asia-Pacific Mussel Watch: monitoring of butyltin contamination in coastal waters of Asian developing countries. Environ. Toxicol. Chem., 21: 2119–2130.

Talmage, S.S. and B.T. Walton, 1991. Small mammals as monitors of environmental contaminants. Rev. Environ. Contam. Toxicol., 119: 47–145.

Tanabe, S., 1994. International mussel watch in Asia-Pacific. Mar. Pollut. Bull., 28: 518.

Tanabe, S., 2000a (Ed.). Mussel Watch: Marine Pollution Monitoring in Asian Waters. Report on the Monbusho Grant-in-Aid for International Scientific Research Program (Field Research) in the fiscal years 1997–1999 (Project No. 09041163), p. 156.

Tanabe, S. 2000b. Asia-Pacific Mussel Watch Progress Report. Mar. Pollut. Bull., 40: 651.

Tanabe, S., R. Tatsukawa and D.J.H. Phillips, 1987. Mussels as bioindicators of PCB pollution: a case study on uptake and release of PCB isomers and congeners in green-lipped mussels (*Perna viridis*) in Hong Kong waters. Environ. Pollut., 47: 41–62.

Tanabe, S., M.S. Prudente, S. Kan-atireklap and A. Subramanian, 2000. Mussel watch: marine pollution monitoring of butyltins and organochlorines in coastal waters of Thailand, Philippines and India. Ocean Coast. Manage, 43: 819–839.

Ueno, D., S. Takahashi, S. Tanabe, K. Ikeda and J. Koyama, 1999. Uptake kinetics of persistent organochlorines in mussels through transplantation experiment. J. Environ. Chem., 9: 369–378 (Abstract in English).

Ueno, D., S. Takahashi, S. Tanabe, K. Ikeda, J. Koyama and H. Yamada, 2001. Variations in organochlorine concentrations in blue mussels associated with season, size and vertical habitat. Nippon Suisan Gakkaishi, 67: 887–893 (Abstract in English).

Squids: Indicators of Pollution by POPs in Open Oceans

Squids are carnivores and have simple food habits. There are 375 squid species in the oceans around the world; none of these migrate over long distances. Squids transfer contaminants through gills by equilibrium partitioning. The concentrations of toxicants in their bodies would reflect the pollution levels in the seawater at the location and time of their collection; thus, they are good bioindicators for measuring ambient levels of pollution by POPs.

Chapter 3: Squids

Introduction

Like clams and oysters, squids are mollusks and belong to a class called cephalopods, which also includes the octopus and cuttlefish. Their shells are not present outside their bodies but inside. There are approximately 375 species of squids spread across seas and oceans around the world. Their size ranges from 1 inch to approximately 60 feet. Squids, which are 10-armed cephalopods, abound the seas and serve as food for many animals, including whales and dolphins.

Cephalopods are carnivorous. They are mostly bottom dwellers, but at night, they rise to the surface for feeding by increasing their buoyancy. They feed on fish, mysids, crustaceans and other mollusks such as bivalves (Kasugai, 2001). Squids breathe through their folded gills. As in other gill-breathing animals, exchange of chemicals may occur in the gills through equilibrium partitioning.

Mollusks have long been known to naturally accumulate metals and organic chemicals to high concentrations. Although pollutant concentrations in gastropods and bivalve mollusks have been subject to considerable research, those in cephalopod mollusks have been relatively poorly investigated. Recently, some cephalopods have been used as biomonitors to study environmental contamination by POPs. This chapter presents a brief review of POPs in cephalopods.

Monitoring Studies

As early as the 1980s, the tendency of measuring POP concentrations in squids was prevalent among researchers, particularly marine mammal scientists measuring OC concentrations in the tissues of cetaceans and pinnepeds, because squids form a considerable portion of the diet of these animals. For example, Aguilar (1983) found that the concentrations of OC pollutants such as DDTs and PCBs in the sperm whales caught in the coastal waters off Northwest Spain differed from those in other cetaceans in the same area because the sperm whales feed mainly on squids and bottom fish. This indicates that the squids alter the transfer loads of POPs to higher trophic levels. In another instance, Kawano et al. (1986) found that the sum of the CHLs (*cis*-chlordane + *trans*-chlordane +

cis-nonachlor + *trans*-nonachlor + oxychlordane) is gradually concentrated with increase in trophic levels, from zooplankton to Dall's porpoise through squid and fish. As also observed in organisms such as Gentoo penguins, Argentinian hake and flying fish caught in sub-Antarctic waters, the squid (*Illex illecebrosus argentius*) also had 24 individual CB congeners and 15 other OC compounds in their muscle samples. The concentrations of these compounds were lower in these organisms than in those caught from the Northeast Atlantic (deBoer and Wester, 1991); this indicates that the concentrations of these compounds in these organisms reflect their levels in the ambient environment. Weisbrod et al. (2000) found that the bioaccumulation of DDE, CHLs and PCBs was substantially high in some stranded odontocetes such as pilot whales and white-sided dolphins as well as in squids because of their proximity to land-based sources of pollution. In comparison with squids, they found considerably lower levels of these compounds in the fish collected from the same areas.

Weisbrod et al. (2001) found disproportionately high concentrations of total PCBs in the long-finned squid collected from the Northwest Atlantic than in the mackerel and herring collected from the same location. They stated that squids could be a significant source of PCBs to odontocetes living and feeding in the Northwest Atlantic.

Tanabe et al. (1984), Kawano et al. (1986) and Ichihashi et al. (2001) have already found that squids accumulate many types of hazardous chemicals such as PCBs and OC pesticides as well as metals. Tanabe et al. (1984, 1987) and Ueno et al. (1999) reported that gill-breathing organisms have the ability to respond rapidly to changes in the ambient levels of POPs; this is based on the concept of equilibrium partitioning between ambient water and body lipids. Various authors (Tanabe et al., 1984; Iwata et al., 1993; Wania and Mackay, 1996; Tanabe, 2000) have reported that compared with the less lipophilic chemicals such as HCHs, OCs such as DDTs and PCBs are highly resistant to degradation and are less transportable to distant locations; hence, they tend to accumulate in sediments. This might have led to the observation by Weisbord et al. (2000) that bottom-dwelling squids from the Northeast Atlantic had higher concentrations of DDE and CHLs than pelagic fishes from the same area. This may have also led to the conclusion that squids can be used as bioindicators of highly lipophilic chemicals that may reach the sea bottom reservoirs.

Fitness of Squids as Bioindicators

Since squids transfer contaminants through the gills by equilibrium partitioning (Tanabe et al., 1987), the PTS concentrations in their bodies would reflect the pollution levels at the location and time of their collection (Ueno et al.,

2003). Yamada et al. (1997) collected 13 species of squids from 77 stations in oceans around the world and quantified the concentrations of PCBs, TBT and triphenyltin (TPT) in their liver samples. They found a clear spatial variation in all the three compounds. Higher concentrations of all the three compounds were found in specimens collected in the Northern Hemisphere than in those collected near the Southern Hemisphere. In particular, the squid specimens collected near Japan, North America and Europe had considerably higher concentrations of these compounds than those collected from the coastal waters of South American countries, Africa and Australia (Fig. 3.1). Moreover, higher concentrations of these compounds were found in specimens collected from the coastal regions of Japan and North America than in those collected from open oceans of the North Pacific. The authors suggested that the interhemispheric differences in the magnitude of concentrations of organotins and PCBs in the squids may be indicative of the recent increase in the use of TBT and TPT in countries located in the Northern Hemisphere. Although PCB usage has been banned since the 1970s, large amounts of PCBs have been used over a long period in the Northern Hemisphere; hence, PCBs still persist in this region. They also found that the distribution pattern of PCBs, TBT and TPT in their specimens reflects the global usage (amount, period and manner of utilization) of these chemicals. These results show the suitability of squids for use as bioindicators of POPs.

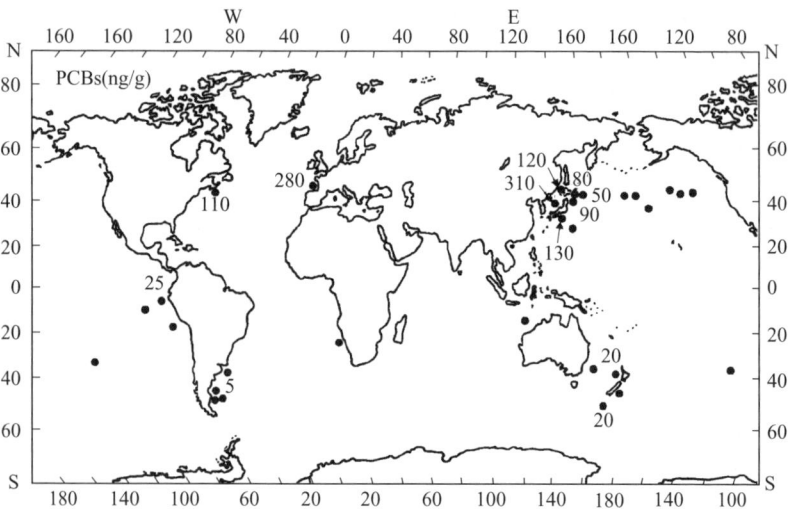

Fig. 3.1. PCB concentrations in squid livers in oceans around the world. Numbers indicate the concentrations (Source: Yamada et al., 1997).

After evaluation of the POP concentrations in tissues, Ueno et al. (2003) reported that the Japanese common squid (*Todarodes pacificus*) can be used as bioindicators of pollution by POPs in Japanese coastal waters. This species is a common food item and is therefore commercially important in the area. Its distribution is limited to the coastal and offshore waters of Japan, but related species are available in oceans around the world (Okutani, 1983). Several authors have previously verified the utility of this species as a bioindicator of different chemicals (Ichihashi et al., 2001). Such characteristics of the Japanese common squid render it a better species for monitoring pollution by POPs in Japanese waters. In fact, this species has also been included in the 'Squid Watch Program' as a bioindicator for monitoring of pollution by PCBs and organotin compounds (Umezu et al., 1993; Yamada et al., 1997). Using the same species, Ueno et al. (2003) conducted further studies for evaluating temporal and spatial variations in POPs and also made a global comparison with related species of squids. In their study, the authors collected specimens of *Todarodes pacificus* from the Sea of Japan and the Pacific coast of Japan and quantified the concentrations of DDTs, CHLs, HCHs, HCB and PCBs in their livers. PCBs were found to be the predominant compounds, reflecting the pattern observed in fishes and other organisms collected from the same localities (Kannan et al., 1995; Prudente et al., 1997; Takahashi et al., 2001; Ueno et al., 2002; de-Britto et al., 2002). In general, the OC concentrations in mussels from the coast of Sea of Japan were higher than in those from the Pacific coast, reflecting the local pollution sources of OCs. Moreover, the slower water exchange in the Sea of Japan, which is due to an enclosed topography and shallower depth, has prevented the dilution of POPs in the area and has hence resulted in the higher OC concentrations in squid livers.

Ueno et al. (2003) did not find any significant relationship between OCs and either body length (17–26 cm) or body weight (89–400 g) of this species; hence, they concluded that different-sized specimens of this species can be used to determine the temporal and spatial variations in the OC levels in their ambient environment. They found higher PCB concentrations in the specimens collected during the spring (June) than in those collected during fall (October); however, no apparent differences were observed in the concentrations of DDTs, HCHs, CHLs and HCB. This was attributed to the fluctuations in PCB concentrations in the seawater on account of the existing sources of PCBs near the collection site.

Further, significantly higher concentrations of PCBs, DDTs and HCHs were found in the specimens collected from the coast of Sea of Japan than in those from the Pacific coast, whereas HCB concentrations were found to be rather uniform (Fig. 3.2). This may be due to the persistent discharge of PCBs from

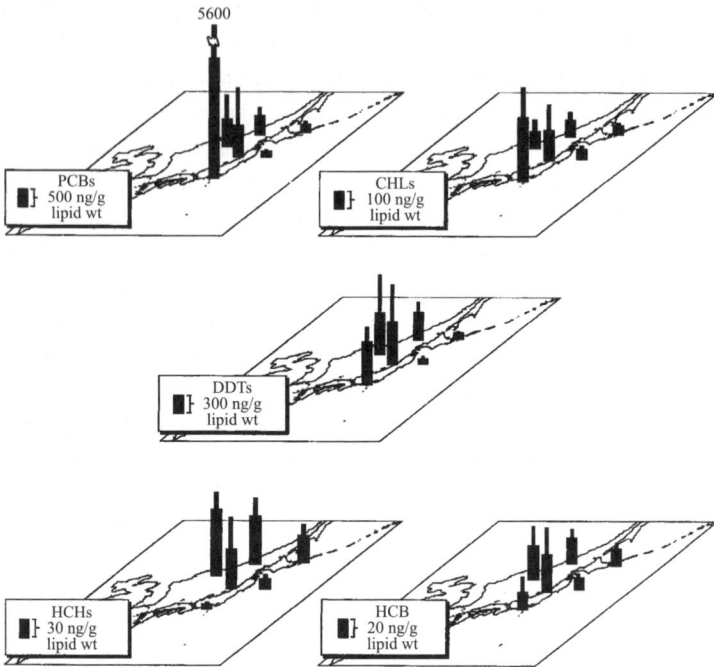

Fig. 3.2. Geographical distributions of OC concentrations in the liver of the Japanese common squid collected from the waters around Japan (Source: Ueno et al., 2003).

Japan and Russia, DDTs and HCHs from China and the former Soviet Union and HCB from almost all the countries bordering the Sea of Japan. All these results suggest that the Japanese common squid rapidly reflects pollution levels in the seawater at the time and location of their collection.

Conclusions

Unlike studies involving other mollusks, particularly bivalves, laboratory and transplantation studies that employ squids for pollution monitoring are not available. However, a few existing reports that provide monitoring data on POPs testify to their utility as possible bioindicator organisms for POP monitoring in shallow coastal environments. From the available literature, it has been found that OC concentrations in some of the squid specimens reflect their levels in the ambient environment, and seasonal- and growth-related factors do not greatly affect the OC concentrations in squid species, as has been observed in fish. Moreover, related species of squids may show the same concentration characteristics. There are approximately 375 species of squids spread across the oceans around the world; hence, selection of suitable bioindicators may be easier. If required, squids may also be used as bioindicators of pollution by POPs.

References

Aguilar, A., 1983. Organochlorine pollution in sperm whales, *Physeter macrocephalus*, from the temperate waters of the eastern North Atlantic. Mar. Pollut. Bull., 14: 349–352.

deBoer, J. and P. Wester, 1991. Chlorobiphenyls and organochlorine pesticides in various sub-Antarctic organisms. Mar. Pollut. Bull., 22: 441–447.

de-Britto, A.P.X., S. Takahashi, D. Ueno, H. Iwata, S. Tanabe and T. Kubodera, 2002. Organochlorine and butyltin residues in deep-sea organisms collected from western North Pacific, off-Tohoku, Japan. Mar. Pollut. Bull., 45: 348–361.

Ichihashi, H., Y. Nakamura, K. Kannan, A. Tsumura and S. Yamasaki, 2001. Multi-elemental concentrations in tissues of Japanese common squid (*Todarodes pacificus*). Arch. Environ. Contam. Toxicol., 41: 483–490.

Iwata, H., S. Tanabe, N. Sakai and R. Tatsukawa, 1993. Distribution of persistent organochlorines in the oceanic air and surface seawater and the role of ocean on their global transport and fate. Environ. Sci. Technol., 27: 1080–1098.

Kannan, K., Y. Yasunaga, H. Iwata, H. Ichihashi, S. Tanabe and R. Tatsukawa, 1995. Concentrations of heavy metals, organochlorines and organotins in horse shoe crab, *Tachypleus tridentatus* from Japanese coastal waters. Arch. Environ. Contam. Toxicol., 28: 40–47.

Kasugai, T., 2001. Feeding behaviour of the Japanese pygmy cuttlefish *Idosepius paradoxus* (Cepahlopoda: Idiosepiidae) in captivity: evidence for external digestion? J. Mar. Biol. Assoc. U.K., 81: 979–981.

Kawano, M., S. Matsushita, T. Inoue, H. Tanaka and R. Tatsukawa, 1986. Biological accumulation of chlordane compounds in marine organisms from the northern North Pacific and Bering Sea. Mar. Pollut. Bull., 17: 512–516.

Okutani, T., 1983. Todarodes pacificus. In: P.R. Boyle (Ed.), Cephalopod Life Cycles, Vol. I, Specific Account, Academic Press, London, pp. 201–214.

Prudente, M., S. Tanabe, M. Watanabe, A. Subramanian, N. Miyazaki, P. Suarez and R. Tatsukawa, 1997. Organochlorine contamination in some odontoceti species from the North Pacific and Indian Ocean. Mar. Environ. Res., 44: 415–427.

Takahashi, S., S. Hayashi, R. Kasai, S. Tanabe and T. Kubodera, 2001. Contamination of deep-sea organisms from Tosa Bay, Japan_ by organochlorine and butyltin compounds. Natl. Sci. Mus. Monogr., 20: 363–380.

Tanabe, S., 2000. Asian developing regions: persistent organic pollutants in the seas. In: C.R.C. Sheppard (Ed.), Seas at the Millennium: An Environmental Evaluation. Elsevier Sciences, Amsterdam, pp. 447–462.

Tanabe, S., H. Tanaka and R. Tatsukawa, 1984. Polychlorobiphenyls, DDT, and hexachlorocyclohexane isomers in the western North Pacific ecosystem. Arch. Environ. Contam.Toxicol., 13: 731–738.

Tanabe, S., R. Tatsukawa and D.J.H. Phillips, 1987. Mussels as bioindicators of PCB pollution: a case study on uptake and release of PCB isomers and congeners in green-lipped mussels (*Perna viridis*) in Hong Kong waters. Environ. Pollut. 47: 41–62.

Ueno, D., H. Iwata, S. Tanabe, K. Ikeda and K. Koyama, 1999. Uptake kinetics of persistent organochlorines in mussels through the transplantation experiment. J. Environ. Chem., 9: 369–378 (in Japanese).

Ueno, D., H. Iwata, S. Tanabe, K. Ikeda, J. Koyama and H. Yamada, 2002. Specific accumulation of persistent organochlorines in bluefin tuna collected from Japanese coastal waters. Mar. Pollut. Bull., 45: 254–261.

Ueno, D., T. Inoue, K. Ikeda, H. Tanaka, H. Yamada and S. Tanabe, 2003. Specific accumulation of polychlorinated biphenyls and organochlorine pesticides in Japanese common squid as a bioindicator. Environ. Pollut., 125: 227–235.

Umezu, T., 1993. Squid watch. Look Japan, December 1993, pp. 30–31.

Wania, F. and D. Mackay, 1996. Tracking the distribution of persistent organic pollutants. Environ. Sci. Technol., 30: 390A–396A.

Wesibrod, A.V., D. Shea, G. Leblac, M. Moore, and J.J. Stegeman, 2000. Organochlorine bioaccumulation and risk for whales in a Northwest Atlantic food web. Mar. Environ. Res., 50: 440–441.

Weisbrod, A.V., D. Shea, M.J. Moore and J.J. Stegeman, 2001. Species, tissue and gender-related organochlorine bioaccumulation in white-sided dolphins, pilot whales and their common prey in the Northeast Atlantic. Mar. Environ. Res., 51: 29–50.

Yamada, H., K. Takayanagi, M. Tateishi, H. Tagata and K. Ikeda, 1997. Organotin compounds and polychlorinated biphenyls of livers in squid collected from coastal waters and open oceans. Environ. Pollut., 96: 217–226.

Fishes Reflect Pollution by POPs in all Water Bodies

Fishes are ubiquitous and occur in almost all water bodies. Most fish species are good accumulators of persistent and lipophilic compounds such as POPs and reflect the environmental levels of these compounds in most instances. Fishes can be used as bioindicators of POP exposure in international waters after careful consideration of their various life history parameters.

Chapter 4: Fish

Introduction

Since aquatic ecosystems are the ultimate reservoirs of many contaminants such as POPs, either due to direct discharge or hydrologic and atmospheric processes, measurement of the levels of bioaccumulated toxins in the aquatic ecosystems may serve as an indirect measure of the pollutant levels in the nearby terrestrial environments. Meanwhile, the biological behaviour of pollutants, such as bioavailability, bioaccumulation and biotransformation, as well as pollution-induced biological and biochemical effects on aquatic organisms have been increasingly used in recent years to evaluate or predict the impact of chemicals on aquatic ecosystems.

Beyer (1996) opined that for assessment of the quality of aquatic ecosystems, all the mandatory bioindicator criteria are met by numerous fish species. Fish are ubiquitous in the aquatic environment and play a major role in the aquatic food webs because they function as carriers of energy from lower to higher trophic levels (Beyer, 1996). Therefore, understanding the uptake and bioaccumulation of POPs in various fish tissues and organs is highly relevant to studies on bioindicators. Many of the general biomarker criteria, for example, changes in enzymes and protein levels, seem to be easily interpretable as due to the specific toxicity of certain specific chemical exposures in fish. Because of such characteristics, fish are considered as one of the best animals to study the accumulation characteristics and effects of chemicals in aquatic ecosystems.

A monitoring species should be selected from an exposed community based on the relationship of the species to the exposure and assessment end point (Suter, 1993). Almost all fish species fulfil these criteria because being aquatic animals, they never leave the water. However, different fish species may exhibit considerable variations in both accumulative (bioindicator response) and basic physiological features (biomarkers), and their specific responsiveness towards environmental pollution may be apparent. Despite limitations such as a relatively high mobility and different feeding habits, fish are generally considered to be one of the most feasible organisms for pollution monitoring of aquatic ecosystems.

Bioaccumulation and biomarker responses in fish may be measured and interpreted in order to elucidate the aquatic behaviour of chemical contaminants. Fish are one of the most suitable bioconcentrators to identify POPs and to assess the exposure of aquatic organisms to POPs (Schmidt et al., 1990; Kannan et al., 1995; Corsi et al., 2003; Van der Oost et al., 2003). Since it is virtually impossible to predict the fate of these xenobiotic substances in a simple partitioning model, the complexity of bioaccumulation should be considered along with bioaccumulation, metabolism and organ-specific concentrations of contaminants.

Bioaccumulation Mechanisms

Rachel Carson's (1962) book 'Silent Spring' ushered an era of research on biomagnification and food chain relationships. The information presented in this book led to a widespread assumption that as a general rule, biomagnification occurs from the lower to higher trophic levels in food webs. This rule may be partially valid, particularly in situations such as the closely knit predator-prey relationship in which a predator is characteristically dependent only on a single or a few prey species. On the contrary, aquatic ecosystems, particularly marine ecosystems, are more open than terrestrial ecosystems. In the marine environment, predators consume a wide range of prey organisms that are smaller in size than the predator.

POPs are highly hydrophobic chemicals that may accumulate in fishes depending on their lipid content. Various theories have been formulated for explaining POP concentrations in fishes. However, scientists face unresolved controversies related to the mechanism of accumulation of pollutants in aquatic animals, particularly gill-breathing animals such as fishes. Moreover, in the case of fishes with different feeding habits, the accumulation pattern of POPs may greatly vary among species.

Knowledge regarding bioaccumulation and the levels of chemicals in biota is a prerequisite to understanding the adverse effects of the chemicals on ecosystems (Franke et al., 1994). POPs may accumulate in fishes via various pathways, for example, via direct uptake from water through gills or skin (bioconcentration), ingestion of particulate matter from water (ingestion) and/or consumption of contaminated food (biomagnification). Normally all these processes occur in varying degrees of combinations in all fishes. Even if acute or chronic effects are not detected in toxicity tests, accumulation of POPs in fish tissues should be regarded as a hazard criterion because some effects may be recognized only at a later stage of life, may be multigenerational (e.g. impact of PCBs on the egg hatching success; Tillitt et al., 1992) or may manifest themselves only in higher members of the food web. Contaminant concentrations in fishes may vary depending on species, age groups, reproductive status and various other parameters.

Contaminat concentrations in fishes are determined primarily by the uptake and elimination kinetics, which are typical for both chemicals and fish species under study. Bioconcentration of POPs in fish depends on the transfer characteristics between water and fish and the excretion and metabolism of these chemicals. Uptake of organic pollutants in fish may occur directly via exchange with water phase or indirectly via the consumption of contaminated food (Thomann, 1989). Although biotransformation of trace organic pollutants has been reported in fish, clearance mainly occurs by simple release from the gill membranes and via fecal excretion into the surrounding water (Brown, 1994).

Factors Affecting Bioaccumulation

Oppurhuizen (1991) stated that the bioconcentration factor (BCF) of a chemical in fish is the ratio of its concentration in the body of the organism to that in water during steady state or equilibrium. For the partitioning of chemicals between water and the lipid phases of organisms, the steady state BCF can be defined as

$$BCF = K_W/K_B = C_B/C_W$$

where C refers to concentration; k, rate constant; and B and W, biota and water.

Yang et al. (2000) demonstrated a significant correlation between the uptake and depuration of OC compounds and oxygen consumption, regardless of fish size and species. They also showed that the uptake rate depends on the OC concentrations in water, which are generally higher for less hydrophobic compounds.

Apart from the hydrophobicity of POPs, the lipid content of fish also plays a major role in bioaccumulation. Spaice and Hamelink (1982) showed that the uptake rate of hydrophobic chemicals in fish increases with an increase in lipid content in their biological membranes. The time required for reaching a steady state of POP concentrations in water and fish can be determined by caging uncontaminated fish in polluted areas and measuring tissue contaminant concentrations after different exposure times. For example, in the rainbow trout, the estimated equilibrium times ranged between 15 and 256 days for various PCB congeners (Vigano et al., 1994) and between 56 and 275 days for persistent pesticides (Galassi et al., 1996).

In large predatory fishes, biomagnification probably plays a more prominent role than bioconcentration. Biomagnification is the ratio of the uptake of chemicals from food to their clearance (Sijm et al., 1992), whereas bioconcentration is the process of uptake of pollutants from the ambient environment by passive diffusion through gills or skin. Since the levels and burdens of persistent and extremely hydrophobic compounds (such as POPs) cannot always be explained satisfactorily by simple partitioning processes between sediment, water and fish (Thomann, 1989), it is likely that uptake via contaminated food (biomagnification) contributes significantly to the accumulation of these contaminants in fish (Van der Oost et al., 2003).

The octanol/water partition coefficient (Kow) of POPs also plays a major role in their accumulation in fish. By comparing predicted levels of a food chain bioaccumulation model with recorded levels in field observations, Thomann (1989) demonstrated that biomagnification was significant for chemicals with log K_{ow} > 5. The observed bioaccumulation levels in feral fish indicated significant elevations (up to two orders of magnitude) above the calculated levels. Leblanc (1995) demonstrated that significant biomagnification was observed for chemicals with log K_{ow} > 6.3. He suggested that trophic level differences in bioaccumulation might be due to increased bioconcentration as a consequence of decreased elimination efficiencies of organisms occupying higher trophic levels.

Fisk et al. (1998) showed that persistent OCs with K_{ow} ≈ 7 have the greatest potential for food chain accumulation in fish. Russell et al. (1999) demonstrated a gradient in the food chain accumulation of persistent chemicals; those with log K_{ow} > 6.3 were highly biomagnified, those with log K_{ow} between 5.5 and 6.3 had an intermediate biomagnification potential and those with log K_{ow} < 5.5 did not show any evidence of biomagnification. On the other hand, Van der Oost et al. (2003) reported that although some linear relationships have been demonstrated to exist between octanol/water partition coefficient and the biomagnification factors, their relationship was very poor in case of many types of chemicals. They further stated that log K_{ow} cannot be expected to be a good indicator of the bioaccumulative behaviour of all organic chemicals in fish because many factors influencing accumulation, for example, the phenomena of active transport, diffusion through cell membranes, metabolism and accumulation in specific organs, are normally not taken into consideration. A fish with all its biochemical constituents, metabolic capacities, enzymatic and hormonal controls, etc. cannot simply be considered as a fat globule in water, which can accumulate lipophilic chemicals present in water depending on the octanol/water partition coefficient.

Gray (2002) opined that high OC concentrations in fish may not be due to biomagnification (uptake through food) because uptake of contaminants in these organisms may occur through their body surface or respiratory organs by diffusion, which is a process of bioconcentration. For most small organisms such as plankton, polychaetes, bivalves and crustaceans, the major route of intake is through respiratory surfaces. Randall et al. (1998) showed that in trout (*Salmo trutta*), the largest proportion of tetrachlorobenzene uptake occurred through the gills. They are of the opinion that intake through food is a relatively minor pathway. Based on all such observations, Gray (2002) concluded that at least with respect to organisms up to the level of fish in the food chain, biomagnification may not be a major pathway for accumulation of persistent chemicals.

Gobas et al. (1993) explained the mechanism of biomagnification and food chain accumulation of organic chemicals by a fugacity-based approach. MacDonald et al. (2002) stated that the three processes—bioconcentration, bioaccumulation and biomagnification—are analogous to the chemical extraction of organic chemicals using solvents and evaporation of solvents. They explained that the bioconcentration of persistent chemicals in fish is similar to solvent switching; these chemicals are transferred from the water to lipid phases in fish. On the other hand, biomagnification is caused by solvent depletion as lipids are digested. Bioaccumulation is the sum of both the processes.

Fugacity is equivalent to chemical activity or chemical potential because it pertains to the tendency of a chemical to escape from a phase such as water or food (Clark et al., 1988). A difference in fugacity provides a driving force for net passive chemical transport from high to low fugacity phases. Based on the fugacity concept, Mackay and Fraser (2000) explained the development of a fugacity gradient in the gut of fish and the resulting chemical diffusion into the cells. Food digestion in the gastrointestinal tract was found to produce a 4- to 5-fold increase in chemical fugacity in the food by altering the fugacity capacity, while an additional 2- to 3-fold increase in the chemical concentration and fugacity is caused by a reduction of the food volume due to absorption from the gastrointestinal tract (Gobas, 1993; Gobas et al., 1993).

Food digestibility and absorption were found to be critical factors controlling biomagnification and dietary uptake efficiencies under laboratory and field conditions (Gobas et al., 1999). Based upon dietary information, bioenergetics modelling and determination of PCBs in both predator and prey fish, Madenjian et al. (1998) estimated that the lake trout and coho salmon in Lake Michigan retained 80% and 50% of the PCBs contained in their food, respectively.

As a result of biomagnification, experimentally derived BCF values may deviate significantly from those predicted with partitioning models (Thomann, 1989; Van der Oost et al., 1996). Biomagnification may be more important in larger fish than in smaller fish because relative gill ventilation volumes decrease with size, while relative feeding rates are almost equal (Opperhuizen, 1991).

When biomagnification is an important uptake route, individual site-specific variations in bioaccumulation patterns may occur due to fish diet (Van der Oost et al., 1996).

Bioavailability of a chemical is the fraction of the total amount of the chemical present in the ambient environment that can be potentially taken up by an animal into its tissues during its lifespan. Bioavailability is another factor that affects the uptake of POPs in fish (Belfroid et al., 1996). While measuring the bioindicator capacity of a fish, failure to consider the actual bioavailable fraction of the relevant POP may result in underestimation of the bioconcentration potential. Moreover, sediment characteristics such as particle size and organic matter content may be important factors that determine the bioavailability of hydrophobic chemicals such as POPs. Usually, ingestion of small sediment particles rich in organic matter can result in increased contaminant uptake.

Belfroid et al. (1995) indicated that sorption and bioavailability of any hydrophobic chemical were affected by its residence time in soil and sediments. According to White et al. (1999), the longer the residence time in sediments, the lower is the bioavailability of certain compounds such as PAHs. Further, it has been noted that in the aqueous phase, bioaccumulation of PCBs and chlorobenzenes (Schrap and Opperhuizen, 1990), pesticides (Muir et al., 1994), PCDDs and PCDFs (Loonen et al., 1994) and PAHs (Haitzer et al., 1999) is affected by the presence of particulate matter such as sediments, humic acids and other organic matter. In these studies, it has been suggested that a reduction in the uptake of hydrophobic chemicals is caused by reduced bioavailability of the compounds due to sorption to particles. On the other hand, sediment-bound chlorobenzenes have been found to be available to benthic deposit feeders, indicating that ingestion of sediment particles is a significant uptake route of these hydrophobic pollutants (Boese et al., 1990).

Biotransformation is another process that may affect the bioaccumulation process of POPs in fish. This process generally leads to the formation of a more hydrophilic metabolite, which is more easily excreted. Biotransformation may also alter the toxicity of a compound; this may be either beneficial or harmful to an organism. This process may thus be an important determinant of the half-life of a compound in the body. Such processes may cause severe deviations in the accumulation potential of POPs.

However, when uptake rates are significantly higher than metabolic clearance rates, bioaccumulation continues even if the chemical is readily biodegradable. Then, the pollutant concentrations in tissues can be a function of factors controlling the activity of biotransformation enzymes, thus affecting the suitability of the organism as a bioindicator.

Several other conditions such as spawning, nutritional status, health, duration of exposure and life-cycle stages affect the POP accumulation and excretion patterns in fish. Synergistic and antagonistic effects of exposure to two or more pollutants may also affect the concentration of pollutants in animal bodies. Such processes of biotransformation of xenobiotic organic chemicals in fish have been extensively reviewed by Sijm and Opperhuizen (1989). Van der Oost et al. (1996) reported that fish tissue concentrations of chemicals that are easily biotransformed (e.g. low-chlorinated PCBs, PAHs, non-2,3,7,8-substituted PCDDs and PCDFs) are most likely not suitable as bioindicators of accumulation because these concentrations do not reflect the levels in the surrounding environment.

Many modelling experiments were conducted on the accumulation of POPs in fish; these experiments were based on the Equilibrium Partitioning Theory (Thomann et al., 1992; Gobas, 1993; Hendricks, 1995; Morrison et al., 1997). In these models, it was assumed that the concentration of a chemical in a fish is solely determined by the chemical concentration in the water phase and the lipid content of the fish species.

Monitoring Studies

Although many studies have been conducted to investigate the accumulation of organic trace pollutants in aquatic organisms, no standardized methodology was used. Since the extremely low levels of contamination in the ambient water column prevent the reliable quantification of contaminants, in many cases, fish have been used for assessing the status of pollution. Since the capacity of fish to accumulate hydrophobic contaminants is highly dependent upon their tissue lipid content, it is important to express the concentration in the tissues both on lipid and wet weight bases for comparative research. Nongovernmental organizations (NGOs) and government organizations have released innumerable scientific publications and reports regarding the concentrations of several POPs in fish. However, very few studies used the data for biomonitoring the status of POPs in the ambient environment.

The following studies are among the important researches in which the authors were successful in using the POP concentrations in fish tissues for assessing the status of contamination in the respective regions. OC concentrations in fish have been measured and extensively reported over the past four decades (Vieth et al., 1977; Falandysz, 1981; Kawano et al., 1985, 1988; de Boer, 1989; Kelly and Campall, 1994; Kannan et al., 1995; Monirith et al., 1999, 2000; Takahashi et al., 2000; de Brito et al., 2002; Ueno et al., 2002, 2003). In most of these studies, the OC concentrations in fish tissues were used to discuss environmental contamination based on the contamination status of the surrounding waters and/or the migratory areas of fish. However, the values reported before 1980 may be regarded as somewhat unreliable because the earlier analytical methods used packed GC columns available at that time.

In the US, the National Contaminant Biomonitoring Program (NCBP), which originated in 1967 as a segment of the National Pesticide Monitoring Program (NPMA), has periodically analyzed residues of selected OC contaminants in fish and wildlife samples collected from a nationwide network of sampling sites. This program can be cited as one of the pioneering studies in which fish were used as bioindicators of pollution by POPs in the US. The data on OCs obtained from freshwater fish collected from 1967 to 1981 were immediately published in the form of various reports (Henderson et al., 1969, 1971, 1972; Walsh et al., 1977; May and MacKinney, 1981; Schmidt et al., 1981, 1983, 1985; Lowe et al., 1985). Although some of the papers may have involved the use of older methods, they provide the basic data on OCs in fish.

Referring to the previous reports, Schmidt et al. (1990) found statistically significant variations in the temporal and spatial trends in several POPs such as DDTs, PCBs, dieldrin, endrin, CHL, toxaphene, mirex and HCB as well as some other organic contaminants such as HCHs. Using composite fish samples collected from 112 stations located at key points in the major rivers in the US during 1984 and 1985, they found that some of the most persistent OC compounds were at lower concentrations than at any time since the early 1970s. The average concentrations of DDTs, PCBs, CHL, dieldrin and toxaphene—the contaminants occurring at the highest concentrations and at most stations— declined significantly when compared with those recorded during 1976 (Fig. 4.1), particularly at the stations with the highest concentrations. This indicated

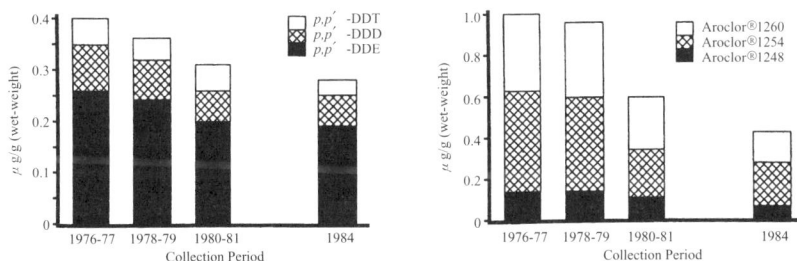

Fig. 4.1. Geometric mean concentrations of p,p'-DDT homologues and PCBs (as Aroclor® mixtures) in fish from U.S. rivers, (1976–1984) (Source: Schmitt et al., 1990).

the suitability of fish as bioindicators for assessing the temporal variations in POPs in aquatic ecosystems.

While acknowledging the usefulness of POP concentrations in fish as bioindicators for monitoring environmental quality, the authors also cautioned about the differences that may occur in the POP concentrations in fish bodies due to the differences in the metabolic capacity of fish for different isomers and congeners of the same compound (e.g. PCBs and toxaphene), which may affect the accumulation patterns. Contrary to the previous as well as present prevalent opinion, the authors believe that lipid normalized residue concentrations of these compounds in fish tissues cannot improve the precision of the data.

Kawai et al. (1988) provided an explanation for such abnormalities arising from the lipid normalization of POP concentrations in fish tissues based on the differential distribution of OCs among different organs and lipid types, which may presumably differ between species and even among individuals of the same species. Hence, Schmidt et al. (1990) recommended the standardization of a common lipid profile determination method among monitoring programs. The authors have also suggested the possibility of using the ratios of the metabolites to the parent contaminant (e.g. DDE/DDT) for determining the chronology of input of such chemicals into the environment.

In an attempt to monitor the contamination by OCs such as DDT and PCBs along the US-Mexico border, Mora et al. (2001) used the contaminant loadings in fish and birds collected in and around the Lower Rio Grande Valley, extending from Texas, USA, to Tamaulipas, Mexico, situated near the Gulf of Mexico. The authors found differences in the extent of DDT contamination in the same fish species collected from both sides of the US-Mexico border. DDE concentrations in the white crappie (*Pomoxis* spp.) from the US were 2.5 times greater than in those from Mexico. Similarly, DDE concentrations in the gizzard shad (*Dorosoma cepedianum*) were 3.3 times greater in the fish from the US than in those from Mexico. In other studies conducted in the same region (TNRCC, 1994, 1997; Davis et al., 1995), DDE concentrations in the fish from the US side were found to be three times to three orders of magnitude higher than in those from Mexico. This clearly demonstrated the suitability of fish for monitoring the spatial variations in POPs. The authors also found that the DDE concentrations in the fish collected by them from the US side were greater than in fish collected during the 1980s and 1990s; this indicates that temporal variations in POPs may be traced using fish as biomonitors.

PCB concentrations in fish from both the sides were below the limits of detection; this reflects the effect of the location of the sampling sites, which were mostly adjacent to agricultural areas. While working on the OC and organotin residues in myctophid fishes from the western North Pacific,

Takahashi et al. (2000) found lower concentrations of PCBs, CHLs and BTs (butyltin compounds) in oceanic species than in fishes from the Suruga Bay along the coast of Japan. Another interesting pattern observed by the authors was the specific trend in the concentrations of these compounds in accordance with the diel vertical migration by these fishes. Relatively high concentrations of PCBs, DDTs and CHLs were found in non-migratory species living in deeper waters, whereas the concentrations of HCHs, HCB and BTs were high in migratory species that migrate to the upper 200 m at night for feeding (Fig. 4.2). The observed results indicate a significant decrease in the input of DDTs into the marine environment, while the slower decline in HCHs indicates its continuous input at least until recently. These patterns were also influenced by the vertical distributions of OCs and BTs in the subarctic and transitional waters of the western North Pacific because of the intrusion of the water mass from the Okhotsk Sea. The authors found wide variations in the OC concentrations

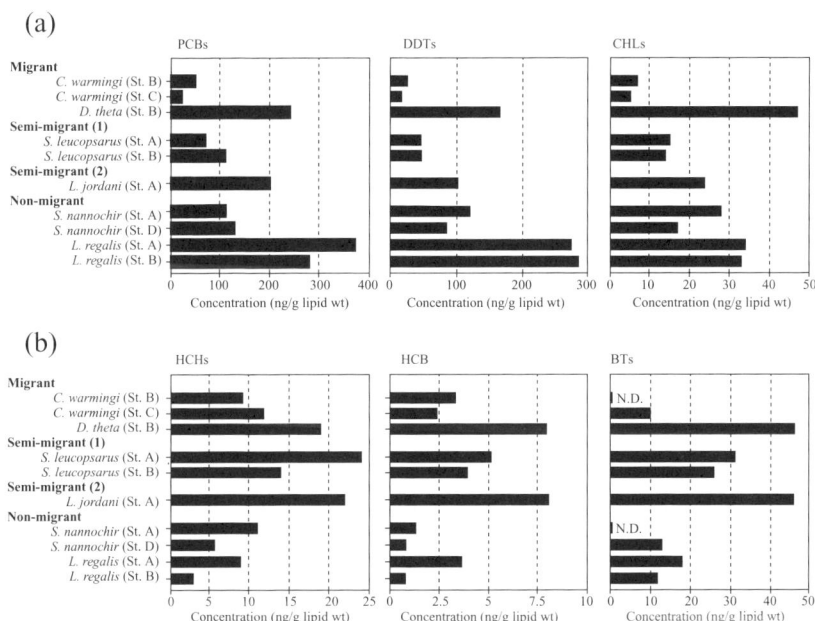

Fig. 4.2. Concentrations and residue patterns of PCBs, DDTs, CHLs, HCHs, HCB and BTs in myctophid fishes with different migration types. The residue patterns were grouped into (a) high concentrations of PCBs, DDTs and CHLs observed in species living in deeper waters and (b) high concentrations of HCHs, HCB and BTs observed in species migrating to shallow waters (Source: Takahashi et al., 2000).

in fish bodies depending on their feeding habits and migration patterns, thus raising doubts regarding the suitability of the use of these fishes for monitoring the status of pollution by POPs in their habitats. By analyzing the composition of OCs, they could trace the chronology of OC usage in that area.

In an effort to use the walleye pollock (*Theragra chalcogramma*) from the Bering Sea, Gulf of Alaska and Sea of Japan as indicator species for measuring contamination by OC compounds, de Brito et al. (2002) showed that specimens from the Bering Sea and Gulf of Alaska showed no differences in the concentrations of PCBs, DDTs, CHLs, HCHs and HCB, whereas significantly higher concentrations (DDTs and HCHs, $p < 0.001$; PCBs, CHLs and HCB, $p < 0.05$) were found in the livers of specimens from the Sea of Japan than in those from the other two areas. By using the values from this indicator species, the authors concluded that there is a large input of DDTs and HCHs from the surrounding countries into the Sea of Japan. Apart from the higher mobility of HCHs via atmosphere and hydrologic processes from countries such as India (Iwata et al., 1993; Wania and Mackay, 1996), the authors also indicated at an overwhelming local input from China and Russia. Using their data on fish, the authors could substantiate the findings of Iwata et al. (1993), that is, HCHs and HCB have a rather uniform distribution in the surface waters of the northern North Pacific, Bering Sea and Gulf of Alaska. They also suggested the possibility of local pollution sources for PCBs and CHLs in the Sea of Japan.

Ueno et al. (2002) used the bluefin tuna (*Thunnus thynnus*) as a bioindicator for monitoring OC pollution; this fish has a habitat extending form the coastal waters to open oceans in both the temperate and tropical regions all over the world. They found that high uptake rates via food and slower equilibration with ambient water and body lipids played a major role in the bioaccumulation of PCBs, DDTs and CHLs in tuna. In contrast, the concentrations of HCB and HCHs were uniform and did not increase linearly with body length, owing to their faster rates of attaining equilibrium between ambient water and body lipids, which was caused by the less lipophilic nature of these compounds. Similar accumulation patterns of these compounds were noted in some other fishes from the Gulf of St. Lawrence and Lake Burtnieku in Latvia (Harding et al., 1997; Olsson et al., 2000). To overcome this problem, Ueno et al. (2002) calculated the body-length normalized value (BLNV) (concentration estimated to 100 cm body length using linear regression analysis) (Fig. 4.3) and found that the values were reflective of the spatial distribution of OCs in the same sampling areas. A similar spatial distribution of OCs was reported in the same study area in sea water (Iwata et al., 1993), fishes (Environmental Agency of Japan, 1997) and marine mammals (Prudente et al., 1997).

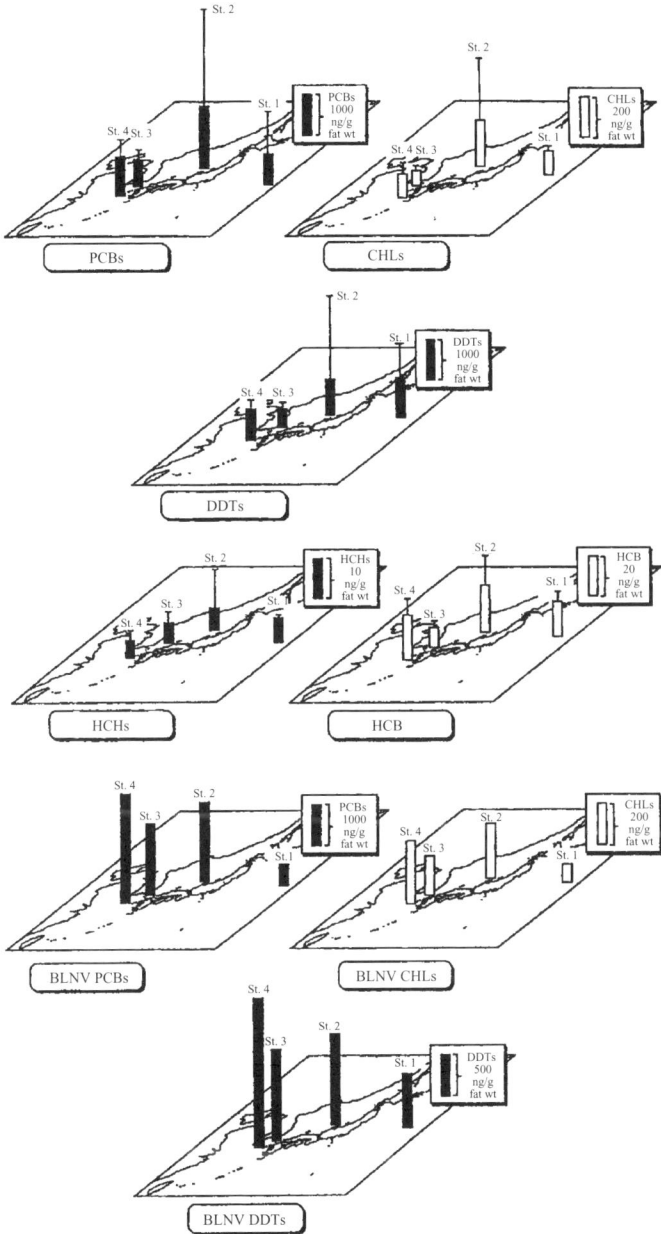

Fig. 4.3. Distribution of measured and body-length normalized value (BLNV) of OC concentrations in the bluefin tuna collected from the coastal waters of Japan (Source: Ueno et al., 2002).

In the case of the skipjack tuna (*Katsuwonus pelamis*), the authors (Ueno et al., 2003) did not find any such discrepancy while using the OC concentrations in fish as bioindicators of DDTs, CHLs, HCHs, HCB and PCBs. The livers of the fish collected from the offshore regions of Japan, Taiwan, Philippines, Indonesia, Seychelles and Brazil as well as the South China Sea, Bay of Bengal and North Pacific showed uniform concentrations of these compounds among individuals within a location. Within a location, no significant difference was observed between growth stage (body length and weight) and OC concentrations. Interestingly, the variations in the OC concentrations in the skipjack tuna could reflect the pollution levels in seawater at the location and time of their collection (Fig. 4.4). The authors found clear spatial variations in the OC concentrations in the fish collected near different countries; this reflected the usage pattern in the nearby countries and the transportability of the chemicals (Fig. 4.5). Based on all these findings, the authors recommended the species *Katsuwonus pelamis* as one of the most suitable bioindicator organisms for global monitoring of POPs.

Most recently, Kajiwara et al. (2003) analyzed the OC concentrations in five species of sturgeons, namely, beluga (*Huso huso*), Russian sturgeon (*Acipenser gueldenstaedtii*), stellate sturgeon (*Acipenser stellatus*), Persian sturgeon (*Acipenser persicus*) and ship sturgeon (*Acipenser nudiventris*), from the Caspian Sea—the biggest land-locked sea—bordered by Russia, Azerbaijan, Kazakhstan, Turkmenistan and Iran. Although considerable gender-dependent variation was observed in the OC concentrations in beluga when compared with the remaining data, the authors concluded that growth- and gender-dependent accumulation are generally less in fish, mollusks and crustaceans. Clayton et al. (1977) and Tanabe et al. (1984) also found such gender-dependent differences in POP concentrations in various fish species. Kajiwara et al. (2003) opined that lower trophic animals such as fish exchange OCs through gills and rapidly equilibrate the OC levels between the ambient water and body lipids. They also suggested that although data from male and female specimens can be pooled for determining spatial and temporal variations in OCs, the effect of other confounding factors, including individual migration, food habits and species specificity of accumulation of different POPs, should be duly considered. Residue concentrations and contamination patterns in sturgeons varied between species and also between their countries of origin. In general, when comparing the residue concentrations from the same species, OC concentrations were found to be the highest in specimens from Azerbaijan and Kazakhstan and lowest in those from Turkmenistan (Fig. 4.6). The authors considered the levels of industrialization and the presence of oil wells in the regions when comparing the status of contamination in the sturgeons.

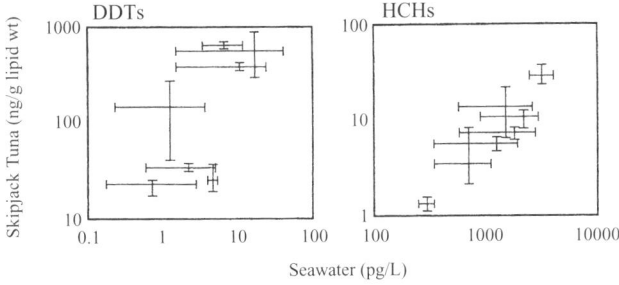

Fig. 4.4. Relationship between DDT and HCH concentrations in the skipjack tuna and surface seawater collected from the same regions. OC concentrations in the surface seawater have been cited from the work of Iwata et al. (1993) (Source: Ueno et al., 2003).

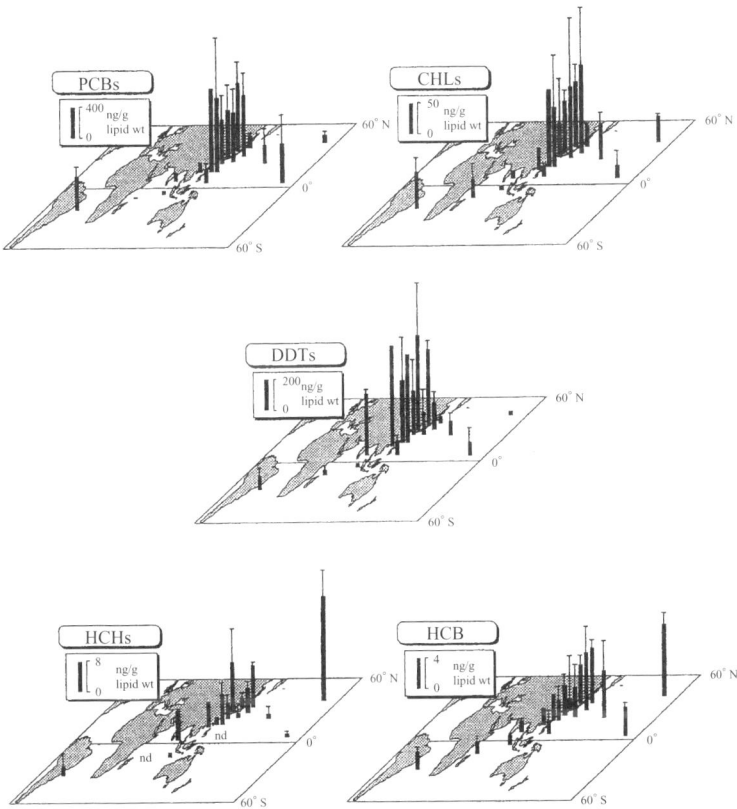

Fig. 4.5. Geographical distributions of OC concentrations in the liver of the skipjack tuna from the waters off Asian countries, Seychelles and Brazil and from open seas (Source: Ueno et al., 2003).

When compared with the innumerable publications reporting successful efforts in using fish as bioindicators of pollution by POPs in the colder regions of the world (e.g. Schmidt et al., 1990; Kannan et al., 1992; Green and Knutzen, 2003; de Brito et al., 2002), only a limited number of successful works are available from the tropics (Kannan et al., 1995; Monirith et al., 2000) where most of the developing countries are located and which are now becoming sources of several POPs.

While comparing the geographical distribution and accumulation features of OC residues in fish from the tropical regions of Asia and Oceania, Kannan et al. (1995) stated that, in general, the OC concentrations were lower in tropical fish than in temperate fish. Residue concentrations in fish showed little spatial variation, as reported for tropical sediments. This is in contrast to the patterns observed for air and water in which higher concentrations were detected in tropical latitudes than in mid-latitudes. As a suitable explanation, the authors suggested that the short residence time of the semi-volatile OCs in the tropical environment due to prevailing high atmospheric temperatures resulted in lower residue concentrations in fish. Following comparison of global data, the authors further suggested the possible utility of the OC concentrations in marine fishes for explaining the spatial variations in the global pollution by these compounds (Fig. 4.7).

In their attempts to monitor the concentrations of OCs, PCBs, DDTs, CHLs, HCB and HCHs in marine and freshwater fishes as well as marine and estuarine bivalves from Cambodia, Monirith et al. (1999, 2000) reported the suitability of several fish species and the green mussel *Perna viridis* for defining the contamination status of the coastal regions of several provinces in Cambodia. By comparing their results with those from global surveys, they concluded that Cambodia has a comparatively 'clean environment' with respect to OC pollution.

Conclusions

Several recent attempts were successful in using different fish species as bioindicators for monitoring pollution by POPs. Fishes are ubiquitous, occurring in almost all water bodies, which are the ultimate reservoirs for all the chemicals used in different terrestrial ecological niches. Most fish species are very good accumulators of persistent and lipophilic compounds such as POPs and reflect the environmental levels of these compounds in most instances. Accumulation of POPs in the body tissues and organs of fish can be used to elucidate the occurrence and behaviour of these compounds in the aquatic environment. While using various fish species as bioindicators of POP levels, it is essential to consider the complexity of the accumulation process, including the metabolism

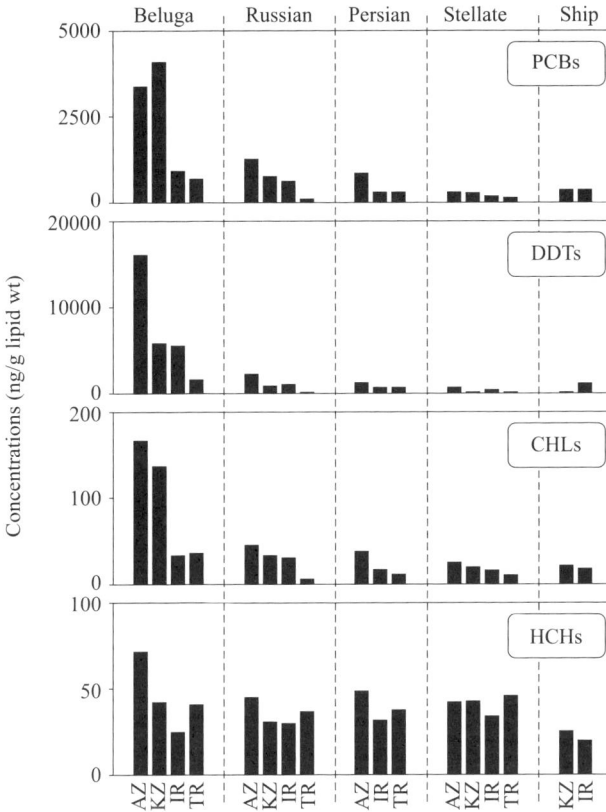

Fig. 4.6. Distribution of OC compounds in sturgeons collected from the Caspian Sea. AZ: Azerbaijan; KZ: Kazakhstan; IR: Iran; TR: Turkmenistan (Source: Kajiwara et al., 2003).

and resulting biotransformation, bioavailability of the chemical, organ specific accumulation, feeding habits and migration pattern.

Although confounding factors such as age, sex, season and mobility may have an additional impact on the bioindicator response, with careful consideration, fish can be used as bioindicators of POP exposure in international waters. Generally, body burdens (muscle, liver or whole body) are the most promising fish bioindicator response for assessment of POP exposure. PCBs, HCB, DDTs, dieldrin, aldrin, endrin and HCHs have been demonstrated to be highly bioaccumulated in fish; hence, the concentrations of these compounds in fish can be easily used for assessment of exposure.

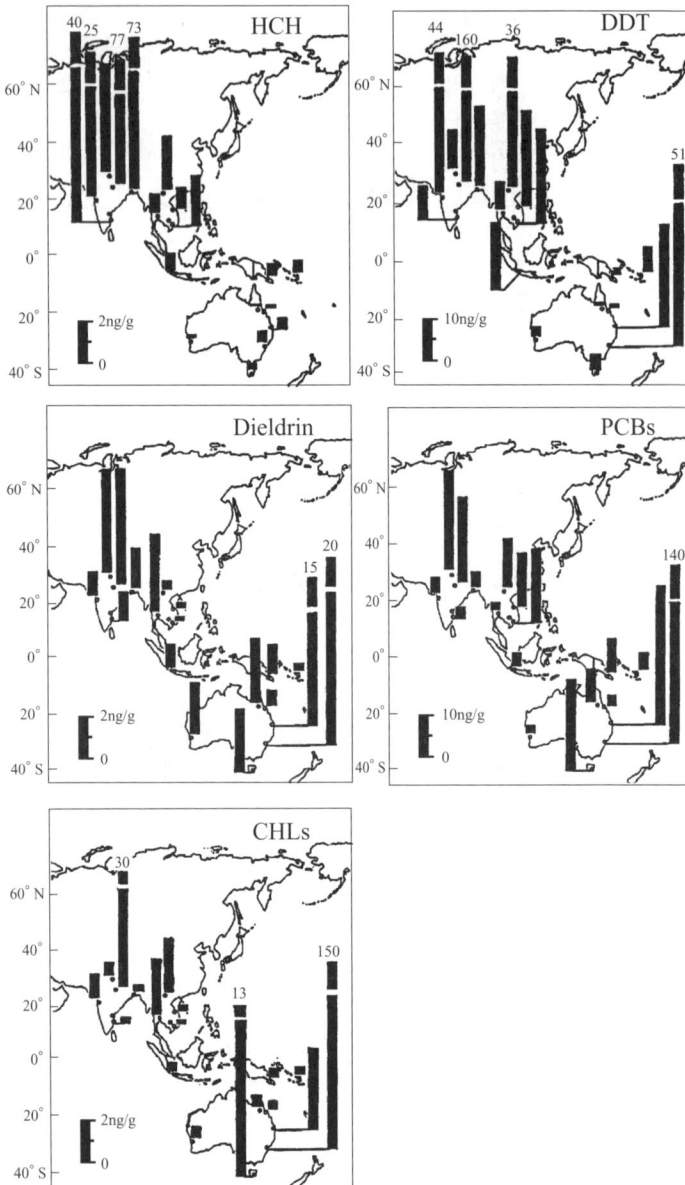

Fig. 4.7. Mean concentrations of OC compounds in fish collected from several locations in eastern and southern Asia and Oceania (Source: Kannan et al., 1995).

To enable comparison of data from global studies, future studies on fish bioindicator response should be carried out using standardized procedures. All the data regarding fish should be expressed both on lipid weight and wet weight bases. Wherever possible, the environmental levels (e.g. sediment levels) of these compounds should also be expressed on the basis of organic matter. Studies involving both tissue-level concentration estimation and toxicity testing in fish are always most reliable and promising; such combination studies should be continued in the future. In other words, future studies should include both bioindicator responses and biomarker studies.

References

Belfroid, A.C., W. Seinen, M. van der Berg, J. Hermens and C.A.M. van Gestel, 1995. Uptake bioavailability and elimination of hydrophobic compounds in earthworms (Eisenia anderi) in filed contaminated soil. Environ. Toxicol. Chem., 14: 605–612.

Belfroid, A.C., D.T.H.M. Sjim and C.A.M. van Gestel, 1996. Bioavailability and toxicokinetics of hydrophobic aromatic compounds in benthic and terrestrial invertebrates. Environ. Rev., 4: 276–299.

Beyer, J., 1996. Fish biomarkers in marine pollution monitoring: evaluation and validation in laboratory and field studies. Academic Thesis, University of Bergen, Norway.

Boese, B.L., H. Lee II, D.T. Specht, R.C. Randall and M.H. Winsor, 1990. Comparison of aqueous and solid-phase uptake for hexachlorobenzene in tellinid clam, Macoma nasuta (Conrad): a mass balance approach. Environ. Toxicol. Chem., 17: 1236–1245.

Brown, J.F. Jr., 1994. Determination of PCB metabolic excretion and accumulation rates for use as indicators of biological response and relative risk. Environ. Sci. Technol., 28: 2295–2305.

Carson, R., 2002. Silent Spring. 40th Anniversary Edition. Houghton Mifflin Company, Boston, p. 365.

Clark, T., K. Clark, S. Paterson, C. Mckay and R.J. Norstrom, 1988. Wildlife monitoring, modelling and fugacity. Environ. Sci. Technol., 22: 120–127.

Clayton, J.R. Jr., S.P. Paulov and N.F. Breitner, 1977. Polychlorinated biphenyls in coastal marine zooplankton: Bioaccumulation by equilibrium partitioning. Environ. Sci. Technol., 11: 676–682.

Corsi, I., M. Mariottini, C. Sensini, L. Lancini and S. Focardi, 2003. Fish as bioindicators of ecosystem health: integrating biomarker responses and target pollutant concentrations. Oceanol. Acta, 26: 129–138.

Davis, J.R., L. Kleinasasser and R. Cantu, 1995. Toxic contaminants survey of the lower Rio Grande, lower Arroyo Colorado and associated coastal waters. Austin, Texas: Texas Nature Resource Conservation Commission.

de Boer, J., 1989. Organochlorine compounds and bromodiphenylethers in liver of Atlantic cod (Gadus morhua) from the North Sea, 1977–1987. Chemosphere, 18: 2131–2140.

de Brito, A.P.X., D. Ueno, S. Takahashi and S. Tanabe, 2002. Contamination by organochlorine compounds in walleye pollock (Theragra chalcogramma) from the Bering Sea, Gulf of Alaska and the Japan Sea. Mar. Pollut. Bull., 44: 172–177.

Environmental Agency of Japan, 1997. Chemicals in the environment. Environmental Health Department, Environment Agency, Japan, p. 568.

Falandysz, J., 1981. Organochlorine pesticides and PCBs in cod-liver oil of Baltic origin. 1971–1980. Pestic. Monit. J., 15: 51–53.

Fisk, A.T., R.J. Norstrom, C.D. Cymbalisty and D.C.G. Muir, 1998. Dietary accumulation and depuration of hydrophobic organochlorines: bio-accumulation parameters and their relationship with the octanol/water partition coefficient. Environ. Toxicol. Chem., 17: 951–961.

Franke, C., G. Studinger, G. Berger, S. Bohling, U. Bruckmann, D. Cohors-Fresenborg and U. Johncke, 1994. The assessment of bioaccumulation. Chemosphere, 29: 1501–1514.

Galassi, S., L. Vigano and M. Sanna, 1996. Bioconcentration of organochlorine pesticides in rainbow trout caged in river Po. Chemosphere, 32: 1729–1739.

Gobas, F.A.P.C., 1993. A model for predicting the bioaccumulation of hydrophobic organic chemicals in aquatic food webs: application to Lake Ontario. Ecol. Model., 69: 1–17.

Gobas, F.A.P.C., X. Zhang and R. Wells, 1993. Gastrointestinal maginification: the mechanism of biomagnification and food chain accumulation of organic chemicals. Environ. Sci. Technol., 27: 2855-2863.

Gobas, F.A.P.C., J.B. Wilcockson, R.W. Russell and G.D. Haffner, 1999. Mechanism of biomagnification in fish under laboratory and field conditions. Environ. Sci. Technol., 33: 133–141.

Gray, J.S., 2002. Biomagnification in marine ecosystems: the perspective of an ecologist. Mar. Pollut. Bull., 45: 46–52.

Green, N.W. and J. Knutzen, 2003. Organohalogens and metals in marine fish and mussels and some relationships to biological variables at reference localities in Norway. Mar. Pollut. Bull., 46: 362–377.

Haitzer, M., S. Hoss, W. Traunpurger and C. Steinberg, 1999. Relationship between concentration of dissolved organic matter (DOM) and the effect of DOM on the bioconcentration of benzo(a)pyrene. Aquat. Toxicol., 45: 147–158.

Harding, G.C., R.J. LeBlanc, W.P. Vass, R.F. Addison, B.T. Hargrave, S. Pearre Jr., A. Dupuis and P.F. Brodie, 1997. Bioaccumulation of polychlorinated biphenyls (PCBs) in the marine pelagic food web, based on seasonal study in the southern Gulf of St. Lawrence, 1976–1977. Mar. Chem., 56: 145–179.

Henderson, C., W.L. Johnson and A. Inglish, 1969. Organochlorine insecticide residues in fish. Pestic. Monit. J., 3: 145–171.

Henderson, C., A. Inglish and W.L. Johnson, 1971. Organochlorine insecticide residues in fish, fall 1969. Pestic. Monit. J., 5: 1–11.

Henderson, C., A. Inglish and W.L. Johnson, 1972. Mercury residues in fish, 1969–1970. Pestic. Monit. J., 6: 144–159.

Hendricks, A.J., 1995. Modelling equilibrium concentrations of micro contaminants in organisms of Rhine delta: can average field residues in the aquatic food chain be predicted from laboratory accumulation. Aquat. Toxicol., 31: 1–25.

Iwata, H., S. Tanabe, N. Sakai and R. Tatsukawa, 1993. Distribution of persistent organochlorines in the oceanic air and surface seawater and the role of ocean on their global transport and fate. Environ. Sci. Technol., 27: 1080–1098.

Kajiwara, N., D. Ueno, I. Monirith, S. Tanabe, M. Pourkazemi and D.G. Aubrey, 2003. Contamination by organochlorine compounds in sturgeons from Caspian Sea during 2001 and 2002. Mar. Pollut. Bull., 46: 741–747.

Kannan, K., J. Falandysz, N. Yamashita, S. Tanabe and R. Tatsukawa, 1992. Temporal trends of organochlorine concentrations in cod-liver oil from the southern Baltic proper, 1971–1989. Mar. Pollut. Bull., 24: 358–363.

Kannan, K., S. Tanabe and R. Tatsukawa, 1995. Geographical distribution and accumulation features of organochlorine residues in fish in tropical Asia and Oceania. Environ. Sci. Technol., 29: 2673–2683.

Kawai, S., M. Fukushima, N. Miyazaki and R. Tatsukawa, 1988. Relationship between lipid composition and organochlorine levels in tissues of dolphin. Mar. Pollut. Bull., 19: 129–133.

Kawano, M., S. Tanabe, T. Inoue and R. Tatsukawa, 1985. Chlordane compounds in the marine atmosphere from the southern hemisphere. Trans. Tokyo Univ. Fish., 6: 59–66.

Kawano, M., S. Matsushita, T. Inoue, H. Tanaka and R. Tatsukawa, 1988. Bioaccumulation and residues patterns of chlordane compounds in marine animals: invertebrates, fish, mammals and sea birds. Environ. Sci. Technol., 22: 792–797.

Kelly, A.G. and A. Campall, 1994. Organochlorine contaminant in liver of cod (Gadus moruha) and muscle of herring (Chupea herrangus) from Scottish waters. Mar. Pollut. Bull., 28: 103–108.

Leblanc, G.A., 1995. Trophic-level differences in the bioconcentration of chemicals: implications in assessing environmental biomagnification. Environ. Sci. Technol., 29: 154–160.

Loonen, H., M. Tonkes, J.R. Parsons and H.A.J. Govers, 1994. Bioconcentrat-ion of polychlorinated dibenzo-p-dioxins and polychlorinated dibenzofurans in guppies after aqueous exposure to a complex PCDD/PCDF mixture: relationship with molecular structure. Aquat. Toxicol., 30: 153–169.

Lowe, T.P., T.W. May, W.G. Brumbaugh and D.A. Kane, 1985. National Contamination Biomonitoring Program – Concentrations of seven elements in freshwater fish, 1978–1981. Arch. Environ. Contam. Toxicol., 14: 363–368.

MacDonald, R., D. Mackay and B. Hickie, 2002. A new approach suggests that phenomena, such as bioconcentration, biomagnification, and bioaccumulation, result from two fundamental processes. Environ. Sci. Technol., 36: 457–462.

Mackay, D. and A. Fraser, 2000. Bioaccumulation of persistent organic chemicals: mechanisms and models. Environ. Pollut., 110: 375–391.

Madenjian, C.P., R.J. Hesselberg, T.J. Desorcie, L.J. Schmidt, R.M. Stedman, R.T. Quintal, L.J. Begnoche and D.R. Passino-Reader, 1998. Estimate of net trophic transfer efficiency of PCBs to Lake Michigan lake trout from their prey. Environ. Sci. Technol., 32: 886–891.

May, T.W. and G.L. MacKinney, 1981. Cadmium, mercury, arsenic and selenium concentrations in freshwater fish, 1976–1977. National Pesticide Monitoring Program. Pestic. Monit. J., 15: 14–38.

Monirith, I., H. Nakata, S. Tanabe and T.S. Tana, 1999. Persistent organochlorine residues in marine and freshwater fish in Cambodia. Mar. Pollut. Bull., 38: 604–612.

Monirith, I., H. Nakata, M. Watanabe, S. Takahashi, S. Tanabe and T.S. Tana, 2000. Organochlorine contamination in fish and mussels from Cambodia and other Asian countries. Water Sci. Technol., 42: 241–252.

Mora, M.A., D. Papoulias, I. Nava and D.R. Buckler, 2001. A comparative assessment of contaminants in fish from four resacas of the Texas, USA–Tamaulipas, Mexico border region. Environ. Intl., 27: 15–20.

Morrison, H.A., F.A.P.C. Gobas, R. Lazar, D.M. Whittle and G.D. Haffner, 1997. Development and verification of a benthic/pelagic food web accumulation model for PCB congeners in western Lake Erie. Environ. Sci. Technol., 31: 3267–3273.

Muir, D.C.G., B.R. Hobden and M.R. Servos, 1994. Bioconcentration of pyrethroid insecticides and DDT in rainbow trout: uptake, depuration and effect of dissolved organic carbon. Aquat. Toxicol., 29: 223–240.

Olsson, A., K. Valters and S. Burreau, 2000. Concentrations of organochlorine substances in relation to fish size and trophic position: a study on perch (Perca fluviatilis L.). Environ. Sci. Technol., 34: 4878–4886.

Opperhuizen, A., 1991. Bioconcentration and biomagnification: is a distinction necessary? In: R. Nagel and R. Loskill (Eds.), Bioaccumulation in Aquatic Systems. VCH Publishers, Weinheim, pp. 67–80.

Prudente, M., S. Tanabe, M. Watanabe, A. Subramanian, N. Miyazaki, P. Suarez and R. Tatsukawa, 1997. Organochlorine contamination in some odontoceti species from the North Pacific and Indian Ocean. Mar. Environ. Res., 44: 415–427.

Randall, D.W., D.S. Connell, R. Yang and R.S.S. Wu, 1998. Concentrations of persistent lipophilic compounds are determined by exchange across the gills and not through the food chain. Chemosphere, 37: 1263–1280.

Russell, R.W., F.A.P.C. Gobbas and G.D. Haffner, 1999. Use of semipermeable membrane devices for studying effects of organic pollutants: comparison of pesticide uptake by semipermeable membrane devices and mussels. Environ. Toxicol. Chem., 17: 1815–1824.

Schmidt, C.J., J.L. Ludke and D. Walsh, 1981. Analysis of variance as a method for examining contaminant residues in fish: National Pesticide Monitoring Program. In: D.R. Branson and K.L. Dixon (Eds.), Aquatic Toxicology and Hazard Assessment, Fourth Conference, American Society for Testing and Materials, Philadelphia, ASTM-STP 737, p. 270.

Schmidt, C.J., M.A. Ribick, J.L. Ludke and T.W. May, 1983. Organochlorine residues in freshwater fish, 1979–1979: National Pesticide Monitoring Program. Pestic. Monit. J., 14: 136–206.

Schmidt, C.J., J.L. Zajicek and M.A. Ribick, 1985. National Pesticide Monitoring Program: residues of pesticides in freshwater fish, 1980–81. Arch. Environ. Contam. Toxicol., 14: 225–260.

Schmidt, C.J., J.L. Zajicek and P.H. Peterman, 1990. National contaminant bio-monitoring program: residues of organochlorines in U.S. freshwater fish, 1976–1984. Arch. Environ. Contam. Toxicol., 19: 748–781.

Schrap, S.M. and A. Opperhuizen, 1990. Relationship between bioavailability and hydrophobicity: reduction of the uptake of organic chemicals by fish due to sorption on particles. Environ. Toxicol. Chem., 9: 715–724.

Sijm, D.T.H.M. and A. Opperhuizen, 1989. Biotransformation of organic chemicals in fish: enzyme activities and reactions. In: O. Hutzinger (Ed.), Handbook of Environmental Chemistry Reactions and Processes, Vol. 2E, Springer, Berlin, pp. 163–235.

Sijm, D.T.H.M., W. Seinen and A. Opperhuizen, 1992. Life-cycle biomagnification study in fish. Environ. Sci. Technol., 26: 2162–2174.

Spaice, A. and J.L. Hamelink, 1982. Alternative models for describing the bioconcentration of organics in fish. Environ. Toxicol. Chem., 1: 309–320.

Suter, G.W. II, 1993. Ecological Risk Assessment. Lewis Publishers, Boca Raton, Florida, USA, p. 538.

Takahashi, S., S. Tanabe and K. Kawaguchi, 2000. Organochlorine and butyltin residues in mesopelagic myctophid fishes from the western North Pacific. Environ. Sci. Technol., 34: 5129–5136.

Tanabe, S., H. Tanaka and R. Tatsukawa, 1984. Polychlorobiphenyls, ΣDDT, and hexachlorocyclohexane isomers in the western North Pacific ecosystem. Arch. Environ. Contam. Toxicol., 13: 731–738.

Thomann, R.V., 1989. Bioaccumulation model of organic chemical distribution in aquatic food chains. Environ. Sci. Technol., 23: 699–707.

Thomann, R.V., J.P. Conolly and T.F. Parkerton, 1992. An equilibrium model of organic chemical accumulation in aquatic food webs with sediment interaction. Environ. Toxicol. Chem., 11: 615–629.

Tillitt, D.E., G.T. Ankley, J.P. Giesey, J.P. Ludwig, H. Kuita-Matsuba, D.V. Weseloh, D.V. Ross, C.A. Bishop, L. Sileo, K.L. Stromborg, J. Larson and T.J. Kubiak, 1992. Polychlorinated biphenyl residues and egg mortality in double-crested cormorants from Great Lakes. Environ. Toxicol. Chem., 11: 1281–1288.

TNRCC, 1994. Texas Natural Resources Commission. Binational study regarding the presence of toxic substances in the Rio Grade/Rio Bravo and its tributaries along the boundary portion between the United States and Mexico. Final Report, Austin, Texas.

TNRCC, 1997. Texas Natural Resources Commission. Second phase of the binational study regarding the presence of toxic substances in the Rio Grade/Rio Bravo and its tributaries along the boundary portion between the United States and Mexico. Final Report, Vol. II, Austin, Texas.

Ueno, D., H. Iwata, S. Tanabe, K. Ikeda, J. Koyama and H. Yamada, 2002. Specific accumulation of persistent organochlorines in bluefin tuna collected from Japanese coastal waters. Mar. Pollut. Bull., 45: 254–261.

Ueno, D., S. Takahashi, H. Tanaka, A.N. Subramanian, G. Fillmann, H. Nakata, P.K.S. Lam, J. Zheng, M. Muchtar, M. Prudente, K.H. Chung and S.

Tanabe, 2003. Global pollution monitoring and organochlorine pesticides using skipjack tuna as a bioindicator. Arch. Environ. Contam. Toxicol., 45: 378–389.

Van der Oost, R., A. Opperhuizen, K. Satumalay, H. Heida and N.P.E. Vermeulen, 1996. Biomonitoring aquatic pollution with feral eel (Anguilla anguilla): I. Bioaccumulation: biota-sediment ratios of PCBs, OCPs, PCDDs and PCDFs. Aquat. Toxicol., 35: 21–46.

Van der Oost, R., J. Beyer and N.P.E. Vermeulen, 2003. Fish bioaccumulation and biomarkers in environmental risk assessment: a review. Environ. Toxicol. Pharmacol., 13: 57–149.

Vieth, G.D., D.W. Kuehl, F.A. Fuglisi, G.E. Glass and J. Eaton, 1977. Residues of PCBs and DDT in the western Lake Superior ecosystem. Arch. Environ. Toxicol., 5: 487–499.

Vigano, L., S. Galassi and A. Arillo, 1994. Bioconcentration of polychlorinated biphenyls (PCBs) in rainbow trout caged in the river Po. Ecotoxicol. Environ. Saf., 28: 287–297.

Walsh, D., B. Berger and J. Bean, 1977. Heavy metal residues in fish, 1971–1973. Pestic. Monit. J., 11: 5–34.

Wania, F. and D. Mackay, 1996. Tracking the distribution of persistent organic pollutants. Environ. Sci. Technol., 30: 390A–396A.

White, J.C., M. Hunter, K. Nam, J.J. Pignatello and M. Alexander, 1999. Correlation between biological and physiological availabilities of phenanthrene in soils and soil humin in ageing experiments. Environ. Toxicol. Chem., 18: 1720–1727.

Yang, R., C. Brauner, V. Thurston, J. Neuman and D.J. Randall, 2000. Relationship between toxicant transfer kinetic processes and fish oxygen consumption. Aquat. Toxicol., 48: 95–108

Birds: "Spies" of Local, Regional and Global Pollution

Birds are useful as bioindicators of POPs in certain cases. While the resident birds may reflect the background pollution in their habitat, migratory birds can act as 'spies' of the pollution levels in their wintering and feeding grounds as well as along their migratory routes. With careful sampling strategies, the global pattern of pollution by POPs can be integrated using birds as bioindicators.

Chapter 5: Birds

Introduction

In recent years, considerable attention has been focused on understanding the transport, fate and distribution of POPs because they are found in wildlife samples worldwide, including pristine remote areas (Auman et al., 1997; Muir et al., 1999; Guruge et al., 2001; Wyk et al., 2001; Kunisue et al., 2002; Tanabe, 2002). Elliott and Harris (2001–2002) reported that DDTs and other contaminants such as PCBs, PCDDs and PCDFs have contributed to the reduced nest success and population decline of bald eagles. They focused precisely on the susceptibility of avian species from the viewpoint of toxicological and bioaccumulative characteristics of such chemicals. As a result, recent years have witnessed a dramatic increase in public concern regarding the state of contamination by POPs. Monitoring of these substances must be started in a wide range of ecosystems as the first step towards finding a solution to the problems posed by POPs on a global scale. In order to monitor POPs and their environmental consequences as well as to assess ecosystem health, effective bioindicators must be employed. In this context, birds are one of the appropriate organisms.

There are innumerable reports on pesticide residues in birds. DDT and/or PCBs have been implicated in serious reproductive impairment in some birds, many of which are now endangered species. Biologists studying declining populations of predatory birds found DDE in bird tissues and eggs; however, they could not identify the exact mechanism by which DDE might cause reproductive failure and population declines (Stickel et al., 1966; Peakall and Kiff, 1988). Similar to marine mammals, fish-eating birds have been adversely affected by POP contamination. Reduced detoxification capacity (Tanabe, 2002) and higher exposure via dietary intake results in the accumulation of high PCB concentrations in some seabirds (Walker et al., 1984). Among the avian species, fish-eating birds show serious teratogenic and reproductive dysfunctions as a result of high PCB accumulation (Tillitt et al., 1992; Yamashita et al., 1993), and most of the effects were generally believed to have been due to dioxin-like planar PCBs (Guruge at al., 2000).

Ratcliffe (1967) discovered that shells of raptor eggs from England weighed 18.9% less than they did before the advent of DDT use; this explained the shell breakage during incubation. Later, Hickey and Anderson (1968) found an inverse correlation between DDE residues in eggs and the shell thickness in peregrines, bald eagles and osprey. However, Parslow and Jefferies (1977) demonstrated that for any bird species, the amount of eggshell thinning is closely and linearly correlated with DDE concentrations in the egg (Fig. 5.1), and not with the concentrations of other pollutants such as dieldrin or PCBs (Furness et al., 1993). The authors also stated that once such a relationship is established, it could be used to monitor the level of DDT pollution in a region.

Some researchers argued that DDE was not responsible for eggshell thinning and population decline in birds (Edwards, 1972; Beatty, 1973), but the recovery of bird populations following the ban of DDT use in USA proved that these criticisms were incorrect (Peakall, 1990). In their book 'Environmental Contaminants in Wildlife—Interpreting Tissue Concentrations', Beyer et al. (1996) identified the effects of OC insecticides, PCBs, heavy metals, dioxins, etc. on wildlife.

However, in the case of birds, the problem was of a different nature. Migratory species of both prey and predatory birds are often exposed to pollutants during their migrations (Ramesh et al., 1992; Tanabe et al., 1998; Muir et al., 1999; Tanabe, 2002; Kunisue et al., 2003). Residue analysis helped to identify the nature and locations of pollutant exposure during migration. For example, Henney et al. (1982) found that peregrine falcons accumulate DDE in their bodies during their annual migration to Latin America. Migratory birds of the Arctic region are the best documented examples of such a contamination. Muir

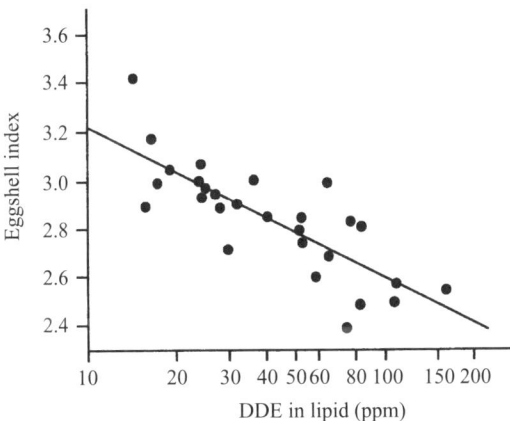

Fig. 5.1. Gannet eggshell thickness index in relation to DDE concentration in lipids in the egg (Source: Parslow and Jefferies, 1977).

et al. (1999) stated that the Arctic birds that breed in the north and over-winter in more industrialized regions contain higher OC concentrations than those that over-winter in the north. Contaminant uptake occurs through food near the over-wintering grounds; these contaminants are transported north each spring when the birds migrate back to their breeding grounds.

Seabird eggs have also been shown to be an efficient, conservative matrix for monitoring OC levels in the marine environment (Gilbertson et al., 1987; Oxynos et al., 1993; Braune et al., 2001). The authors compared the concentrations of various OCs such as DDE, PCB, HCB and oxychlordane in the eggs of different birds found in the Arctic Circle and found a declining trend in the concentrations of these chemicals from the mid-1970s to the late 1980s. In fact, as stated earlier, the toxic effects of POPs on birds were first detected in bird eggs (Moore and Ratcliffe, 1962). Later, Peakal (1974) extracted remnant lipids from museum eggs and showed that DDE was present in peregrine eggs collected as early as 1948.

Unlike in the case of developed countries (Pain et al., 1999; Guruge et al., 2000, 2001; Sakamoto et al., 2002), limited data is available regarding pollutant loads in birds from the developing countries of Asia, South America and Africa (Goldstein et al., 1996; Lacher et al., 1997; Senthilkumar et al., 1998; Tanabe et al., 1998; Wyk et al., 2001; Minh et al., 2002; Kunisue et al., 2003; Sethuraman and Subramanian, 2003). Hence, collection of data on the resident and migratory birds from developing nations has become necessary. At the same time, the possibility of using bird species as bioindicators of pollution by POPs in respective regions may be greatly hampered due to several reasons such as their high mobility, easy susceptibility to pollutants and excretion via moulting and egg laying.

Factors Affecting Accumulation

Migration

Migration is a common phenomenon among birds. The life of birds is governed by activities such as finding food and avoiding predators. They have to face drastically different environmental conditions in summer and winter. Birds are migratory. Almost all birds have at least a small range of migratory territory (Alerstam, 1993).

Avian species are useful bioindicators for monitoring OC contamination in the environment because they are often at a higher trophic level in the food chain. Resident birds, which primarily have localized feeding and breeding grounds throughout the year, may reflect the background pollution in their habitats (Kunisue et al., 2003). Henny and Blus (1986) showed that individuals of the black-crowned herons reproducing at Ruby Lake, Nevada, contained different concentrations of pesticide residues depending upon the different wintering grounds they visited. Springer et al. (1984) also found that the contaminant profiles (relative amounts of different compounds) of populations of peregrine falcons helped to evaluate the origin of residues in falcons and did not reflect the contaminant levels in the ambient environment of the collection area.

In their work on persistent OC residues in resident and migratory birds from Asia, Minh et al. (2002) and Kunisue et al. (2003) found considerable variations in the concentrations of DDTs, PCBs, HCHs, CHLs and HCB between the resident and migrant species collected from the same location. Using the data on the residue concentrations in the resident birds from India, Japan, the Philippines, Russia (Lake Baikal) and Vietnam, Kunisue et al. (2003) suggested the predominant contaminants in each country. However, the migrant birds from different countries showed different patterns of OC residues, indicating that each species has inherent migratory routes and is thus exposed to different contaminants. While recommending great tit (*Parus minor*) nestlings as biomonitors of OC pollution, Dauwe et al. (2003) cautioned that the contamination levels at their natal and breeding sites are different; this may result in differences in the body burdens among breeding females.

Feeding Habits

Birds have a variety of feeding habits. Accordingly, they are classified as herbivores, which feed on terrestrial and marine plants; insectivores; fish-eaters, which forage on the sea bottom; seed-eaters and omnivores (Alerstam, 1993). The pollutant concentrations in their body tissues depend on those in the food organisms. This has been demonstrated in many terrestrial and aquatic bird species.

The avian POP levels depend on their feeding habits. Birds at higher trophic levels may be at greater risk. For example, among the terrestrial birds, insectivores may be at greater risk than non-insectivores (Gard et al., 1995), and among aquatic birds, fish-eaters may accumulate substantial pesticide burdens (Kunisue et al., 2002); this may thus lead to large variations and discrepancies in the interpretation of the monitoring data.

In the 10 species of birds collected by them from the Chubu region of Japan, Hoshi et al. (1998) found that OC contamination was higher in the livers of fish-eating birds and raptors than in those of omnivorous birds. They also found that the ratios of lower chlorinated PCB congeners (tri to tetra) were higher in fish-eating birds than in other birds; this may reflect the effects of feeding habits on drug metabolizing enzyme systems among species.

Olafsdottir et al. (1995) reported high age-dependent concentrations of OCs, including PCBs and DDTs, in the Icelandic gyrfalcon *Falco rusticolus*, a top resident predator. Interestingly, very low OC concentrations were reported in the ptarmigan *Lagopus mutus*, which is by far the most important species in the gyrfalcon diet (Nielson and Cade, 1990). Therefore, in a further study, Olafsdottir et al. (2001) collected six potential prey species, both resident and migratory, in order to elucidate the most likely routes of OC uptake in the gyrfalcon. They found that the OC concentration in the ptarmigan, the primary prey species of the gyrfalcon, was less when compared with those in mallards, golden plovers, purple sandpipers and black guillemots—each of these species accumulates OCs, reflecting the different food chain levels in Iceland. The authors have concluded that although gyrfalcons prey primarily on ptarmigans, they receive a substantial part of their OC load from other marine and migratory birds of the area, which constitute only a minor part of their diet. Therefore, the body burdens of OCs in the predator species of birds are related to the complexity of food chains in different ecosystems, further superimposed by the migratory status of the prey species.

Dauwe et al. (2003) found that insectivorous passerines such as the great tit (*Parus minor*), which has a limited home range, are suitable bioindicators for monitoring terrestrial OC contamination because these birds do not show any age-dependent variation in their feeding habits; hence, they do not exhibit any age-related differences in OC uptake. At the same time, the authors have cautioned regarding the transfer of considerable amounts of OCs from mother to nestling via egg as a result of intensive insect feeding before breeding.

Lemmetyinen et al. (1982), Scharenberg (1991) and Herrera et al. (2000) have documented age-related changes in diet resulting in variations in the intake and body loads of OCs in Arctic terns, cormorants and partridges, respectively; these variations may lead to confusion in interpreting the contamination status of the ambient environment. While acknowledging the possible use of the cormorant *Phalacrocorax carbo sinensis* as a suitable bioindicator in evaluating the OC pesticide levels in Greek wetlands, Konstantinou et al. (2000) also cautioned regarding the large spatial differences in OC concentrations among individuals of the same species; these differences were attributed to the differences in the cormorant's diet between areas and diverse regimes of pollutant management in different wetlands.

Fish-eating birds have been regularly monitored as the species that is potentially at risk of secondary poisoning by POPs. They are top predators in the aquatic food chain and usually accumulate POPs to relatively high internal concentrations. Many authors have found that because of the high concentrations of POPs, particularly PCBs, fish-eating birds can be used as sentinel organisms for monitoring these compounds (Gruge et al., 2001; Jenssen et al., 2001; Kunisue et al., 2002).

Tanabe et al. (1998) observed very clear differences in OC loads in the birds collected from South India; piscivores had higher OC concentrations than omnivores and granivores (Table 5.1).

Table 5.1. Mean concentrations (ng/g wet wt) of OCs in resident and migratory birds according to food habits (Source: Tanabe et al., 1998).

Groups	n	PCBs	DDTs	HCHs	CHLs	HCB
Strict resident						
Inland piscivore and scavenger (BK, LE, PH)	4	37	2,100	3,200	1.7	1.8
Coastal piscivore (CK, WK)	2	100	350	370	0.3	0.3
Omnivore (BD, CM, HC)	7	32	120	590	0.2	0.3
Granivore/occasionally insectvore (CT, MH, SD)	5	<20	130	70	0.1	0.1
Local migrant						
Inland piscivore (BS)	1	30	510	4,100	0.6	2.0
Costal piscivore (KP, LR)	10	180	2,300	750	7.2	0.6
Short-distance migrant						
Inland/coastal piscivore (CR, LP, SP)	17	170	310	240	5.2	5.3
Costal piscivore (WT)	5	2,700	1,000	84	2.6	1.8
Long-distance migrant						
Inland/coastal piscivore (LT, WW)	10	430	680	200	3.9	1.1
Insectivore/piscivore (CS, CA, TS)	13	190	530	280	1.2	0.6

BK = black-winged kite, LE = little egret, PH = pond heron, CK = crested kingfisher, WK = whitebreasted kingfisher, BD = black drongo, CM = common myna, HC = house crow, CT = cotton teal, MH = moorhen, SD = spotted dove, BS = black-winged stilt, KP = kentish plover, LR = little ringed plover, CR = common redshank, LP = long-billed Mongolian plover, SP = short-billed Mongolian plover, WT = white-cheeked tern, LT = lesser-crested tern, WW = white-winged tern, CS = common sandpiper, CA = curlew sandpiper, TS = terek sandpiper

Recently, Tanabe et al. (2004) examined the residue concentrations of PCDDs, PCDFs and coplanar PCBs in albatrosses from open ocean environments. They found that the residue concentrations in albatrosses from the remote North Pacific, far from the point source of pollution, were comparable or higher than those in terrestrial and coastal birds from contaminated areas in developed nations. This suggested specific exposure and accumulation of PCDDs, PCDFs and coplanar PCBs in those birds. Apart from the long lifespan of the species, the authors found that ingestion of plastic and resin pellets by albatrosses could also be a possible reason for the elevated accumulation of these compounds in these birds.

All these factors suggest that the consumption of different types of foods should also be considered while interpreting the OC data in birds.

Metabolism

Minh et al. (2002) evaluated the accumulation profiles of different PCB isomers in the resident and migratory bird species of North Vietnam by calculating the metabolic index (MI) by using the formula proposed by Tanabe et al. (1988):

$$MI_i = \log (CR_{180}/CR_i)$$

where MI_i is the metabolic index of PCB isomer i, CR_{180} is the concentration ratio of CB-180 in the bird and diet and CR_i is the concentration ratio of the congener i. The metabolism of PCB congeners is mediated by CYP-dependent mixed-function oxygenase enzymes such as phenobarbital (PB)- and 3-methylchlonthrene (MC)-type enzymes. The authors compared both the PB- and MC-type enzyme activities in Vietnamese birds with those in birds from other areas and found great variations in the enzyme activities between the different areas (Fig. 5.2). The PB-type enzyme activities were found to be higher in fish-eating birds such as the black-capped kingfisher and whiskered tern; this explains their higher capacity to metabolize POPs and hence the relatively low concentrations of PCBs and other OCs in their tissues.

While discussing the ratios of different toxaphene compounds in the eggs and tissues of adelie penguins and skuas, Vetter et al. (2001) found that nonachlorobornane, one of the higher chlorinated congeners, was highly accumulated in the blubber and brain than in the liver and kidneys, which are the organs of potential metabolism.

Thyen at al. (2000) analyzed PCBs, DDTs, HCB and HCHs in the eggs and chicks of little terns (*Sterna albifrons*) at their breeding colonies in Germany. The PCB composition in the eggs was found to vary depending upon the total

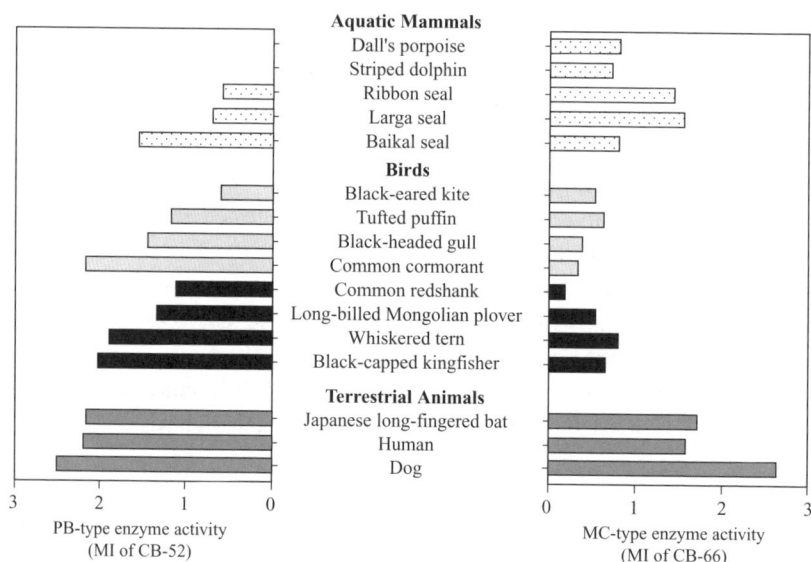

Fig. 5.2. Comparison of estimated phenobarbital (PB)- and 3-methylcholanthrene (MC)-type enzyme activities in higher trophic animals based on the metabolic indices (MI) of chlorobiphenyl (CB)-52 and CB-66. Black bars represent enzyme activities in Vietnamese birds (Source: Minh et al., 2002).

PCB concentrations, suggesting a concentration-dependent metabolization of PCBs in this species. Relatively large variations in the OC concentrations were also observed between different matrices that were analyzed; this indicates metabolization of lower chlorinated PCBs, HCH isomers and HCB during embryogenesis. This type of induced metabolism of PCBs by embryos and newly hatched chicks of several birds was also reported by several authors (Parkinsons and Safe, 1987; Schwarz and Stalling, 1991; Dietrich et al., 1997). Substantial degradation or elimination of HCB from the chick body during embryogenesis has been observed in birds (Vetter et al., 2001).

While reviewing the contamination and toxic effects of persistent endocrine disrupters in marine mammals and birds, Tanabe (2002) found elevated contamination by OCs in open sea animals such as marine mammals and albatrosses. These elevated concentrations can be attributed to their low capacity to metabolize PTS. Hence, aquatic birds and marine mammals accumulate the dioxin-like compounds to much higher concentrations than humans. This implies that open sea animals and birds are at a greater risk from exposure.

The induction of CYP1A1, which is a highly sensitive biomarker for dioxin-like compounds, has been ascertained by measuring the ethoxyresorufin-O-deethylase (EROD) activities in birds (Tillitt et al., 1991; Elliott et al., 1996). Guruge and Tanabe (1997) found a significant correlation between the EROD activities in the hepatic microsomal fraction of Lake Biwa cormorants and total TEQs, which indicates the induction of drug-metabolizing enzymes by PCBs. In a further study, Guruge et al. (2000) also found significant induction of drug-metabolizing enzymes by increased PCB concentrations in birds collected from the Shinobazu Pond in Japan. They also found a correlation between the ratio of IUPAC 169/126 concentrations and total PCB concentrations, which indicates the possibility of induction of P450 enzyme activities in North Pacific albatrosses (Guruge et al., 2001); this was also suggested by Kannan et al. 1993) in marine mammals.

These observations suggest that birds, which are at higher trophic levels, have various well developed drug-metabolizing enzyme systems that are activated by various POPs at different levels. Moreover, the occurrence and induction of such enzyme systems are species specific and also depend on parameters such as feeding habits and migration. Appropriate corrective measures should be taken during the sampling and interpretation of data if and when birds are used as bioindicators of pollution by POPs.

Reproduction and Gender Differences

Guruge et al. (2000) found an age-related increase in POPs in the bodies of common cormorants. However, Guruge et al. (2001) did not find any age- or gender-related accumulation of POPs in large birds such as black-footed albatrosses and Layson albatrosses of the North Pacific and Southern Ocean. Their specimens belonged to an older age class; hence, the specimens could have accumulated high PCB concentrations through long-term exposure. No gender-specific difference in PCB accumulation was observed in these birds. This may be due to a low excretory rate of OCs through egg laying in large sea birds, which was attributed to small egg weight (Tanabe et al., 1986). Further, the authors suggested that since most albatrosses lay only one egg, the age-related accumulation in larger seabirds could be considered for the assessment of contaminant status.

Tanabe et al. (1998) found pronounced gender differences in the concentra-tions and burdens of OCs in the whole-body homogenates of birds collected from South India; the authors found clearly lower concentrations of OCs in females than in males (Table 5.2).

There are differences in the information pertaining to gender difference in OC concentrations in birds. While a few studies reported the presence of elevated OC concentrations in males (Larsson and Lindegren, 1987; Duursma et al., 1989), several reports indicated comparable or greater concentrations in females (Norstrom et al., 1976; Lemmetyinen et al., 1982; Elliott and Shutt, 1993). Kallenborn et al. (1998) also suggested the excretion of lipophilic pollutants from mother to egg in dippers (*Cinclus cinclus* L.) from southern Norway.

All these studies discussed the male-female differences with regard to pollutant concentrations in different body tissues and whole body burdens. However, reduction in body burdens in females is possible through egg laying. A definitive conclusion should be reached using a larger number of species and

Table 5.2. Mean concentrations and burdens of OCs in the whole-body homo-genates of male and female birds (Source: Tanabe et al., 1998).

Species	Sex	n	Fat content (%)	BW-F (g)	HBW (g)	Concentration (ng/g wet wt)			Burden (µg)		
						PCBs	DDTs	HCHs	PCBs	DDTs	HCHs
Local migrant											
Little ringed plover	M	3	7.4	26.3	28.5	310	6,600	1,300	8.2	170	34
	F	2	7.8	29.0	31.7	55	930	600	1.6	27	17
Short-distance migrant											
Common redshank	M	3	10.9	93.9	102.3	120	750	65	11	70	6.1
	F	2	11.1	98.5	107.4	52	380	37	5.1	37	3.6
White-cheeked tern	M	3	6.9	85.3	95.6	3,600	1,200	110	310	100	9.4
	F	2	6.4	77.1	91.6	1,300	710	43	100	55	3.3
Long-distance migrant											
White-winged tern	M	3	12.7	31.4	35.6	700	1,100	270	22	35	8.5
	F	2	11.5	27.2	31.4	330	1,500	510	8.9	41	14

BW-F = body weight without feather. HBW = whole-body weight including feather.

specimens in the future. Until then, care should be taken while using birds as bioindicators of POP contamination.

Tissue Specific Accumulation

Several organs and tissues as well as whole eggs of birds were used for evaluating the contaminant burdens in birds. Kunisue et al. (2003) used the breast-muscle, liver and whole-body homogenates of different birds collected from India, the Philippines and Russia for analyzing PCBs, DDTs, CHLs, HCB and HCHs. The concentration values of contaminants were expressed on a lipid-normalized basis for interpretation of data Kallenborn et al. (1998) analyzed nine egg samples and three bird samples of the bird species *Cinclus cinclus* L. and found that the sum concentrations of DDTs and PCBs in liver and egg samples account for more than 90% of the total burden. Moreover, the sum concentrations of contaminants showed higher average values in eggs than in liver samples. They also found different accumulation patterns of pollutants in egg and liver samples; these differences were attributed to the differences in matrix composition, particularly lipid content. Moreover, Lemmetyinen et al. (1982) as well as Braune and Norstrom (1989) found that PCB concentrations in eggs were usually comparable to those in the pectoral muscle of parent birds.

While evaluating the seasonal fluctuations in OC concentrations in common eider (*Somateria mollissima*) in Iceland, Olafsdottir et al. (1998) found seasonal variations in the OC concentrations in the breast muscle and liver; they attributed this variation to the utilization of fat reserves for energy in winter. Wyk et al. (2001) collected liver, heart, kidney, pectoral muscle, whole blood and clotted blood (since they represent major body systems) from three species of vultures from different locations in South Africa for measuring 14 OCs, including HCH, CHL, dieldrin, endosulfan and heptachlor epoxide. The authors selected these organs because the heart and blood perform the circulatory functions, muscle represents the dominant body mass, liver is the chief site of metabolism and kidneys are the site of excretion of these contaminants. The authors observed statistical differences in the concentration ranges of all the five toxicants in the whole blood, liver, kidney, heart and muscle. However, the authors reported all their values on a wet weight basis and did not report the lipid content of the tissues. Normalizing the values on a lipid weight basis might have reduced these differences to a greater extent. In an elaborate study on Indian birds, Tanabe et al. (1998) indicated that the male-female differences in OCs on a whole body basis should be used to provide an accurate picture on the excretion of OCs via eggs.

Connell et al. (2003) evaluated DDTs, PCBs, HCHs and CHLs in the eggs of two Ardeid species, namely, the little egret (*Egretta garzetta*) and the black-crowned night heron (*Nycticorax nycticorax*), from the egretries in Hong Kong to determine the exposure-associated risk parameter. The eggs can be easily collected without much harm to the population. However, the whole body burdens of the contaminants may give a fairly accurate picture of the contamination levels, at least in the case of small birds. In larger birds, the most suitable sample may be the breast muscle, rather than organs such as the liver and kidneys, which are involved in the metabolism and detoxification of the contaminants. Tissue-specific differences in the accumulation of PCBs, DDE, PCDDs and PCDFs have been demonstrated in bald eagles (Senthilkumar et al., 2002).

Monitoring Studies

Several available publications explain the spatial and temporal distributions of POPs in birds, but most of these publications explain the differences in POPs based on the migratory patterns of birds (Klemens et al., 2000; Minh et al., 2002; Kunisue et al., 2002). Apart from the problems that may arise due to above-mentioned reasons, birds are undoubtedly good bioindicators of pollution by POPs. They have been used worldwide as bioindicators of pollution in their habitats as well as along their migratory routes.

Muir et al. (1999) reviewed the spatial and temporal trends and the effects of contaminants in the Canadian Arctic marine ecosystem and provided a geographic coverage of information on contaminants such as persistent OCs and heavy metals in tissues of marine mammals and sea birds. They found declining concentrations of PCBs and DDT-related compounds in sea birds from the 1970s to 1980s, followed by a levelling off during the 1980s and early 1990s. With respect to other OCs such as CHL, HCH and toxaphene, using the limited data from 1980s to early 1990s, the authors found few significant declines in the OC concentrations in marine mammals and sea birds. The authors found a lack of spatial trends in OCs in sea birds of the Canadian Arctic. The use of different feeding grounds and wintering areas of the birds were cited as possible explanations for these differences in contaminant concentrations in birds. Apart from this, the review provides elaborate accounts of the geographic coverage of the contamination status of the Canadian Arctic marine ecosystem.

Aurigi et al. (2000) used the eggs of aquatic birds such as mallard, greylag goose, mute swan, coot, glossy ibis, spoonbill, little egret, night heron, grey heron, great white egret, red necked grebe, Dalmatian pelican, pygmy cormorant and common cormorant for evaluation of the OC contamination status of the Danube delta in Europe. While comparing the samples collected in 1982 and 1997, the authors found that the contaminant levels were lower in the 1997 samples than in the 1982 samples, with the exception of DDT in mallard. Therefore, they concluded that between 1982 and 1997, overall differences in chlorinated hydrocarbon levels appear to have declined in the eggs of certain resident or short-distant migrant bird species that breed in the Danube delta area.

The eggs and chicks of little terns (*Sterna albifrons*) at different breeding colonies near the Western Baltic Sea were analyzed annually between 1987 and 1996 by Thyen et al. (2000), and the concentrations of PCBs, DDTs, HCB and HCHs were compared. In both types of samples, the concentrations of contaminants were found to decrease over the 9-year period, indicating a continual decrease in the release of these compounds. They also found remarkable variation among individuals as well as between breeding seasons; these variations were unrelated to temporal trends. Based on the pollution patterns in migratory little terns and resident common terns in their common breeding grounds in the Baltic Sea, the authors showed geographic variations in the concentrations of the contaminants in the Baltic Sea and sub-regions.

Similarly, between 1975 and 1988, a gradual decrease was observed in the concentrations of PCBs, DDTs and total chlorobenzenes in the eggs of three seabird species, namely, thick-billed murres (*Uria lomvia*), northern fulmars (*Fulmarus glacialis*) and black-legged kittiwakes (*Rissa tridactyla*), from the Canadian Arctic (Fig. 5.3) (Braune et al., 2001). However, the decline was not noticed in the case of CHL, dieldrin and mirex in the former two species of birds. The authors suggested that the observed temporal trends are due to the changing deposition patterns of xenobiotic compounds. Konstantinou et al. (2000) also found great spatial variations in PCB concentrations in cormorants (*Phalacrocorax carbo sinensis*) collected from four Greek wetlands, but the differences in OC pesticides were smaller. They attributed this to the differences in the cormorants' diet and different regimes of pollutant management in the sampling areas.

Gruge et al. (2001) observed that PCB concentrations in albatrosses from the Southern Ocean were two to three orders of magnitude lower than in individuals from the North Pacific; this reflects the lower exposure of the former to PCBs via prey such as squid and fish, which were also reported to have lower levels of contamination (Yamada et al., 1997). At the same time, the authors indicated that birds from the North Pacific, which live far from pollution sources, were contaminated by PCBs, similar to most coastal and terrestrial fish-eating birds. Such instances reveal that unknown factors such as trophic status and non-point pollution sources, which can produce geographic variations in remote oceans, are yet to be resolved using suitable bioindicator species at different trophic levels as well as environmental samples.

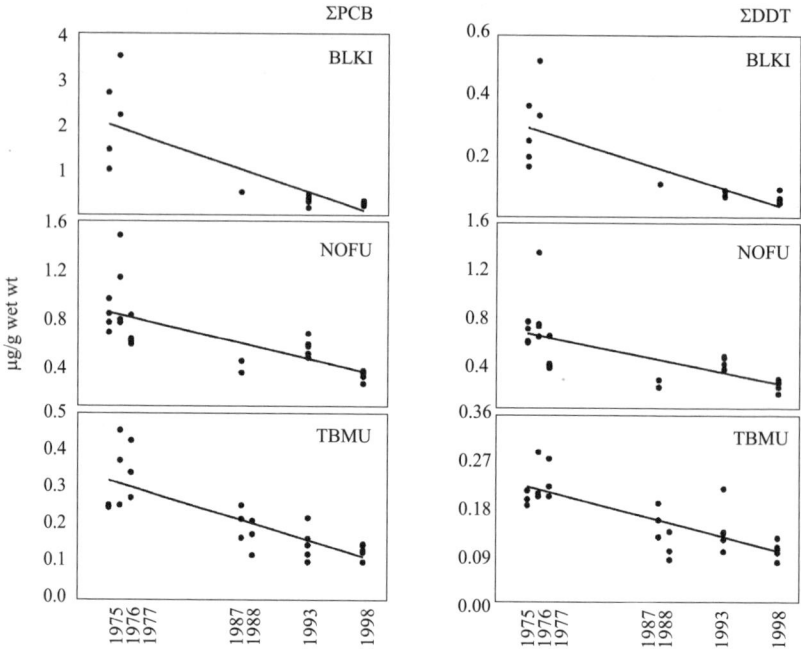

Fig. 5.3. PCB and DDT concentrations (µg/g wet wt) in the eggs of black-legged kittiwakes (BLKI), northern fulmars (NOFU) and thick-billed murres (TBMU) collected between 1975 and 1998. Each point represents the concentration for a pool of three eggs (Source: Braune et al., 2001).

Recently, Tanabe et al. (2004) evaluated the concentrations of PCDDs, PCDFs and coplanar PCBs in five albatross species from the North Pacific and the Southern Ocean for assessing the north-south differences in residue levels. They found that the black-footed and Laysan albatrosses from the North Pacific contained apparently higher concentrations of PCDDs, PCDFs and coplanar PCBs than those from the Southern Ocean, indicating that the emission sources of contaminants were predominant in the Northern Hemisphere. Residue concentrations in albatrosses from the North Pacific, far from the point source of pollution, were comparable or higher than those in terrestrial and coastal birds from contaminated areas of developed nations, suggesting the specific exposure and accumulation of PCDDs, PCDFs and coplanar PCBs in albatrosses. The authors found that the relative proportions of PCDFs and coplanar PCBs in albatrosses from the North Pacific were higher than those observed in birds inhabiting terrestrial and coastal areas; this suggests that compared with

PCDDs, PCDFs and coplanar PCBs may have higher transportability via air and water (Fig. 5.4). Earlier, Auman et al. (1997) noticed significant differences in the total concentrations of PCBs, DDT and DDE in albatross plasma between Layson and black-footed albatrosses collected from remote areas of the North Pacific and different sampling periods. They attributed this to the differences in the mobilization of fat reserves resulting from extended periods of egg incubation without feeding and foraging over great distances to obtain food for the chicks.

By analyzing the contaminant patterns in 400 resident and migratory birds from India, Japan, the Philippines, Russia (Lake Baikal) and Vietnam, Kunisue et al. (2003) found the following spatial differences in contaminant patterns: Japan, PCBs; the Philippines, PCBs and CHLs; India, HCHs and DDTs; Vietnam, DDTs and Russia (Lake Baikal), PCBs and DDTs. By calculating the mean relative concentrations of DDTs, HCHs, CHLs and HCB, which are the ratios of the individual OC concentrations to that of PCBs, the authors could delineate the relative pollution levels of these contaminants in various

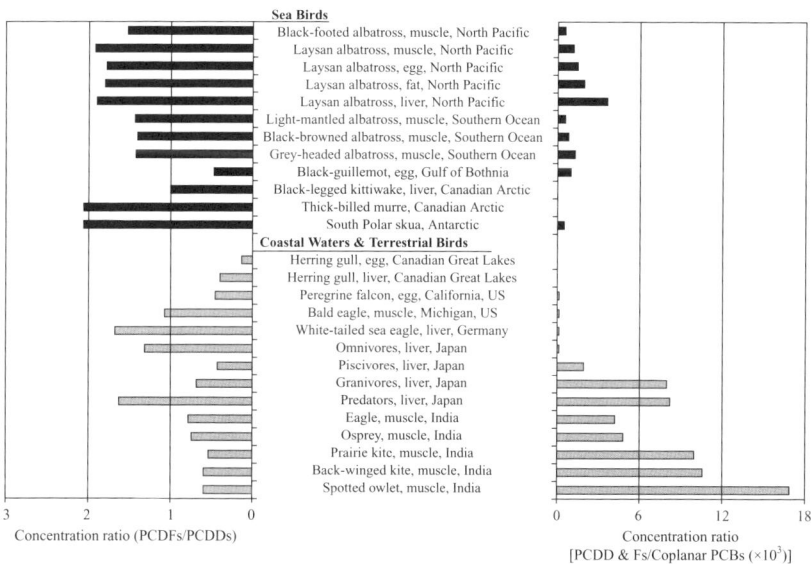

Fig. 5.4. Comparison of the concentration ratios of PCDFs to PCDDs as well as total PCDDs and PCDFs to coplanar PCBs in albatross from the North Pacific and Southern Ocean with those in birds from other locations in the world. The titles indicate the name of species, analyzed tissues and locations (Source: Tanabe et al., 2004).

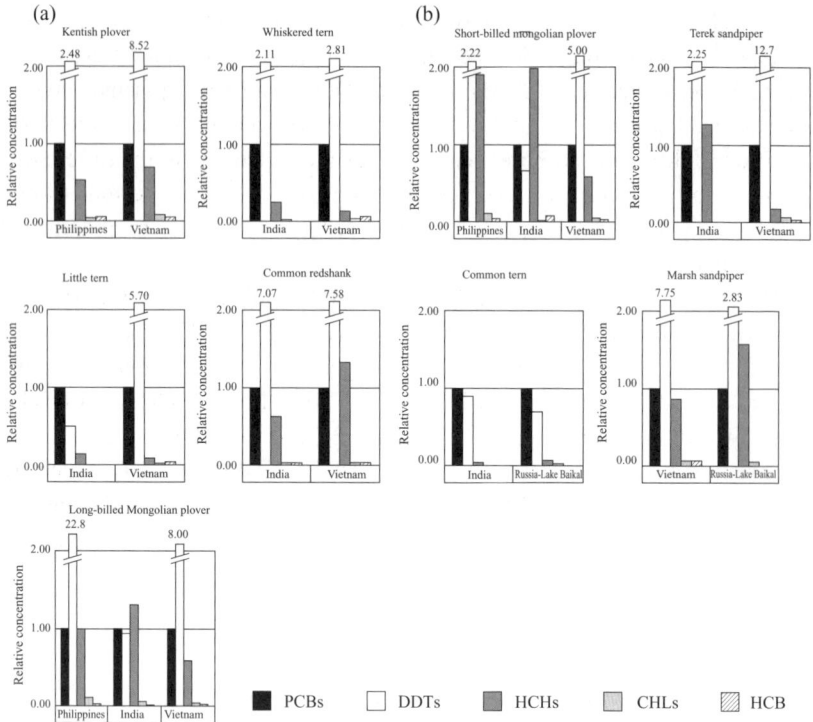

Fig. 5.5. (a) OC residue patterns in same species among short-distance migratory birds collected from Philippines, Vietnam, India and Russia (Lake Baikal). Relative concentration indicates ratio of individual OC concentration to that of PCBs, which was considered as 1.0; (b) OC residue patterns in the same species among long-distance migratory birds collected from Philippines, Vietnam, India and Russia (Lake Baikal). Relative concentration indicates ratio of individual OC concentration to that of PCBs, which was considered as 1.0 (Source: Kunisue et al., 2003).

countries (Fig. 5.5). They also inferred possible migration routes of these birds based on the concentrations of these pollutants in their bodies (Fig. 5.6). The authors further categorized the birds as resident, short-distance migrants and long-distance migrants; they also discussed the different patterns of POP contamination in their bodies based on their routes of migration, species-specific accumulation, feeding habits, etc. By citing several examples of resident and short-distance migrant birds from different countries, the authors showed that these birds can be used as bioindicators of local and regional pollution by POPs. For details on the use of birds as bioindicators of POPs pollution, the readers are referred to these publications and reviews by the research group.

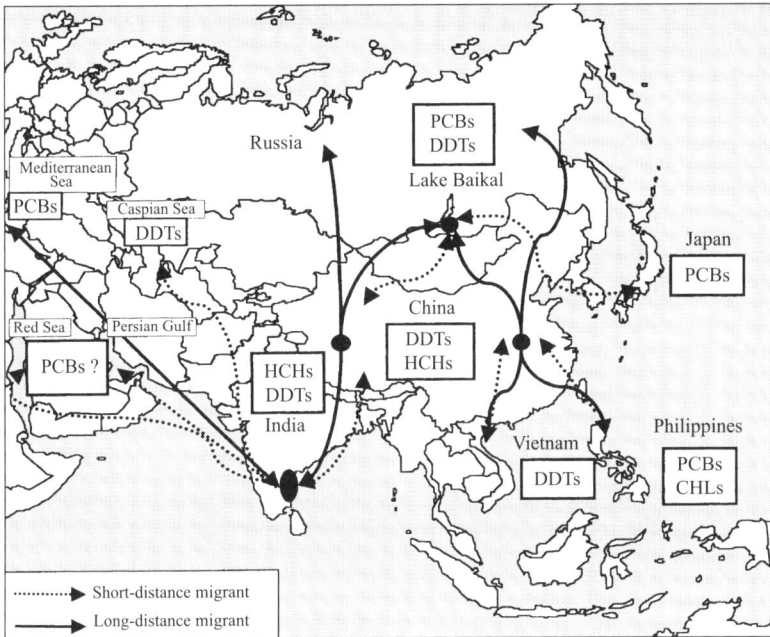

Fig. 5.6. Migratory patterns predicted from accumulation features of OCs in migratory birds collected from Philippines, Vietnam, India and Russia (Lake Baikal). OCs in boxes represent dominant contaminants (Source: Kunisue et al., 2003).

Conclusions

The detection of POP residues in resident and migratory birds clearly showed the environmental problems caused by these chemicals on a global scale. Available studies demonstrated that these contaminants are accumulated in birds in their wintering and breeding grounds and stopover places. Thus, it is evident that birds can definitely be used as biomonitors of pollution by POPs worldwide. In fact, the data on residues in birds (particularly migrants) provided the impetus for raising concern regarding the global transport and distribution of POPs in wildlife. Several recent studies have shown that birds can be used as bioindicators of pollution by POPs on a global scale. However, there are reservations regarding their use on account of their different positions in the food chain, metabolic capacities for POPs, geographical distribution, gender differences, migration and ethical and legal problems pertaining to the collection and transport of the specimens. If and when birds are to be used as bioindicators of POPs, these parameters should be taken into consideration.

Furness et al. (1993) have listed a number of advantages and disadvantages of using birds as biomonitors. They are easy to identify, and their classification and systematics are well established; therefore, there is little risk of monitoring being confounded by uncertainties regarding identities of species or the relationship between the species being studied. Birds are particularly well-studied organisms; considerable research has been conducted on their biology and behavior. This background knowledge of bird biology enhances their usefulness as biomonitors, particularly by reducing the risk of misinterpretations.

Birds are positioned high in the food chain; thus, they may also be sensitive to many diverse factors affecting the food chain. Because of their long lifespan, birds can accumulate the effects of environmental stresses over time, enabling measurements, for example, over a year or more. It may be difficult to establish any short-term perturbations in the environmental changes.

Similarly, the mobility of birds can allow monitoring over a broad spatial scale. However, migratory habits can make birds much less suitable as bioindicators because individuals may differ in their migration patterns to a certain extent and make it difficult to determine the spatial scale they represent. Furthermore, mobility can affect temporal variation; populations of different origins can pass through the same place at different times of the year, thereby potentially confounding a monitoring program based at a particular sampling site.

Birds may be able to regulate tissue concentrations of many chemicals and fat reserves to a much greater extent than invertebrates; therefore, birds may less rapidly reflect changes in the levels of chemicals in the environment, leading to a conclusion that a sedentary invertebrate would be a much better biomonitor than a bird. Although this holds true in many situations, birds also have many advantages as biomonitors of pollution by POPs.

References

Alerstam, T., 1993. Bird Migration. Cambridge University Press, Cambridge, USA, p. 420.

Auman, H.J., J.P. Ludwig, C.L. Summer, D.A. Verburugge, K.L. Froese, T. Colborn and J.P. Giesey, 1997. PCBs, DDE, DDT and TCDD-EQ in two species of albatross on Sand Island, Midway Atoll, North Pacific Ocean. Environ. Toxicol. Chem., 16: 498–504.

Aurigi, S., S. Focardi, D. Hulea and A. Renzoni, 2000. Organochlorine contamination in bird's eggs from the Danube delta. Environ. Pollut., 109: 61–67.

Beatty, B.G., 1973. The DDT myth. John Day, New York, p. 201.

Beyer, W., G.H. Heinz and A.W. Redmon-Norwood, 1996. Environmental Contaminants in Wildlife. A Special Publication of SETAC. CWC Lewis Publishers, Boca Raton, New York, London, Tokyo, p. 494.

Braune, B.M. and R.J. Norstrom, 1989. Dynamics of organochlorine compounds in herring gulls: III. Tissue distribution and bioaccumulation in Lake Ontario gulls. Environ. Toxicol. Chem., 8: 957–968.

Braune, B.M., G.M. Donaldson and K.A. Hobson, 2001. Contaminant residues in seabird eggs from the Canadian Arctic. Part I. Temporal trends 1975–1998. Environ. Pollut., 114: 39–54.

Connell., D.W., C.N. Fung, T.B. Minh, S. Tanabe, P.K.S. Lam, B.S.F. Wong, M.H.W. Lam, L.C. Wong, R.S.S. Wu and B.J. Richardson, 2003. Risk to breeding success of fish-eating Ardeids due to persistent organic contaminants in Hong Kong: evidence from organochlorine compounds in eggs. Water Res., 37: 459–467.

Dauwe, T., S.G. Chu, A. Covaci, P. Schepens and M. Eens, 2003. Great tit (Parus minor) nestlings as biomonitors of organochlorine pollution. Arch. Environ. Contam. Toxicol., 44: 89–96.

Dietrich, S., A. Buthe, E. Denker and H. Hotker, 1997. Organochlorines in eggs and food organisms of avocets (Recurvirostra avosetta). Bull. Environ. Contam. Toxicol., 58: 219–226.

Duursma, E.K., J. Nieuwenhuize, J. van Liere and M.T.H.J. Hillebrand, 1989. Partitioning of organochlorines between water, particulate matter and some organisms in estuarine and marine ecosystems of The Netherlands. Neth. J. Sea Res., 20: 239–251.

Edwards, J.C., 1972. Cracking the thin shell myth. Agric. Chem. Commer. Fert. 27: 20–26.

Elliott, J.E. and M.L. Harris, 2001/2002. An ecotoxicological assessment of chlorinated hydrocarbon effects on bald eagle populations. Rev. Toxicol., 4: 1–60.

Elliott, J.E. and L. Shutt, 1993. Monitoring organochlorines in blood of sharp-shinned hawks (Accipiter striatus) migrating through the Great Lakes. Environ. Toxicol. Chem., 12: 241–250.

Elliott, J.E., R.J. Norstrom, A. Lorenzen, L.E. Hart, H. Philibert, S.W. Kennedy, J.J. Stegeman, G.D. Bellward and K.M. Cheng, 1996. Biological effects of polychlorinated dibenzo-p-dioxins, dibenzofurans, and biphenyls in bald eagle (Haliaeetus leucocephalus) chicks. Environ. Toxicol. Chem., 15: 782–793.

Furness, R.W., J.J.D. Greenwood and P.J. Jarvis, 1993. Can birds be used to monitor the environment? In: R.W. Furness and J.J.D. Greenwood (Eds.),

Birds as Monitors of Environmental Change, Chapman & Hall, London, pp. 1–42.

Gard, N.W., M.J. Hooper and R.S. Bennett, 1995. An assessment of potential hazards of pesticides and environmental contaminants. In: T.E. Martin and D.M. Finch (Eds.), Status and Management of Neotropical Migratory Birds, U.S. Wildl. Serv. Gen. Tech. Rep. RM-229, pp. 310–314.

Gilbertson, M., J.E. Elliott and D.B. Peakall, 1987. Seabirds as indicators of marine pollution. ICBP Tech. Publ., 6: 231–248.

Goldstein, M.I., B. Woodbridge, M.E. Zaccagnini, S.B. Canavelli and A. Lanusse, 1996. Assessment and mortality of Swainson's hawks in Argentina. J. Raptor Res., 30: 106–107.

Guruge, K.S. and S. Tanabe, 1997. Congener specific accumulation and toxic assessment of polychlorinated biphenyls in common cormorants, *Phalacrocorax carbo*, from Lake Biwa, Japan. Environ. Pollut., 96: 425–433.

Guruge, K.S., S., Tanabe and M. Fukuda, 2000. Toxic assessment of PCBs by the 2,3,7,8-tetrachlorodibenzo-*p*-dioxin equivalent in common cormorant (*Phalacrocorax carbo*) from Japan. Arch. Environ. Contam. Toxicol., 38: 509–521.

Guruge, K., M. Watanabe, H. Tanaka and S. Tanabe, 2001. Accumulation status of persistent organochlorines in albatrosses from the North Pacific and the Southern Ocean. Environ. Pollut., 114: 389–398.

Henny, C.J. and L.J. Blus, 1986. Radiotelemetry locates wintering grounds of DDE-contaminated black-crowned night-herons. Wildl. Soc. Bull., 14: 236–241.

Henney, C.J., F.P. Ward, K.E. Riddle and R.M. Prouty, 1982. Migratory peregrine falcons, *Falcus peregrinus*, accumulate pesticides in Latin America. Can. Field Nat., 96: 333–338.

Herrera, A., A. Arino, M.P. Conchello, R. Lazaro, R. Bayarri, C. Yague, J.M. Peiro, S. Aranda and M.D. Simon, 2000. Red-legged partridges (*Alectoris rufa*) as bioindicators for persistent chlorinated chemicals in Spain. Arch. Environ. Contam. Toxicol., 38: 114–120.

Hickey, J.J. and D.W. Anderson, 1968. Chlorinated hydrocarbons and egg shell changes in raptorial and fish eating birds. Science (Wash. DC), 162: 271–273.

Hoshi, H., N. Minamoto, H. Iwata, H. Shiraki, R. Tatsukawa, S. Tanabe, S. Fujita, K. Hirai and T. Kinjo, 1998. Organochlorine pesticides and polychlorinated biphenyl congeners in wild terrestrial mammals and birds from Chubu region, Japan: interspecies comparison of the residue levels and compositions. Chemosphere, 36: 3211–3221.

Jenssen, B.M., V.H. Nielssen, K.M. Murvoll and J.U. Skaare, 2001. PCBs, TEQs and plasma retinol in grey heron (*Ardea cinerea*) hatchlings from two rookeries in Norway. Chemosphere, 44: 483–489.

Kallenborn, R., S. Planting, J. Haugen and S. Nybe, 1998. Congener-, isomer- and enantomer-specific distribution of organochlorines in dippers (*Cinclus cinclus* L.) from southern Norway. Chemosphere, 37: 2489–2499.

Kannan, K., J. Falandysz, S. Tanabe and R. Tatsukawa, 1993. Persistent organochlorines in harbour porpoises from Puck Bay, Poland. Mar. Pollut. Bull., 26: 162–165.

Klemens, J.A., R.G. Harper, J.A. Frick, A.P. Capparella, H.B. Richardson and M.J. Coffey, 2000. Patterns of organochlorine pesticide contamination in neotropical migrant passerines in relation to diet and winter habitat. Chemosphere, 41: 1107–1113.

Konstantinou, I.K., V. Goutner and T.A. Albanis, 2000. The incidence of polychlorinated biphenyl and organochlorine pesticide residues in the eggs of the cormorant (*Phalacrocorax carbo sinensis*): an evaluation of the situation in four Greek wetlands of international importance. Sci. Total Environ., 257: 61–79.

Kunisue, T., T.B. Minh, K. Fukuda, M. Watanabe, S. Tanabe and A.M. Titenko, 2002. Seasonal variation of persistent organochlorine accumulation in birds from Lake Baikal, Russia, and the role of the south Asian region as a source of pollution for wintering migrants. Environ. Sci. Technol., 36: 1396–1404.

Kunisue, T., M. Watanabe, A.N. Subramanian, A. Sethuraman, A. Titenko, V. Qui, M. Prudente and S. Tanabe, 2003. Accumulation features of persistent organochlorines in resident and migratory birds from Asia. Environ. Pollut., 125: 157–172.

Lacher, T.E. and M.I. Goldstein, 1997. Tropical ecotoxicology. Status and needs. Environ. Toxicol. Chem., 16: 100–111.

Larsson, P. and A. Lindegren, 1987. Animals need not be killed to reveal their body burdens of chlorinated hydrocarbons. Environ. Pollut., 45: 73–78.

Lemmetyinen, R., P. Rantamaki and A. Karlin, 1982. Levels of DDT and PCBs in different stages of life cycle of the Arctic tern, *Sterna paradisaea* and the herring gull, *Larus argentatus*. Chemosphere, 11: 1059–1068.

Minh, T.B., T. Kunisue, N.T.H. Yen, M. Watanabe, S. Tanabe, N.D. Hue and N.D.V. Qui, 2002. Persistent organochlorine residues and their bioaccumulation profiles in resident and migratory birds from north Vietnam. Environ. Toxicol. Chem., 21: 2108–2118.

Moore, N.W. and D.A. Ratcliffe, 1962. Chlorinated hydrocarbon residues in the egg of a peregrine falcon (*Falco peregrinus*) from Perthshire. Bird Study, 9: 242–244.

Muir, D.C.G., B. Braune, B. DeMarch, R. Norstrom, R. Wagemann, L. Lockhart, B. Hargrave, D. Bright, R. Addison, J. Payne and K. Reimer, 1999. Spatial and temporal trends and effects of contaminants in the Canadian Arctic marine ecosystem: a review. Sci. Total Environ., 230: 83–144.

Norstrom, R.J., R.W. Risebrough and C.J. Cartwright, 1976. Elimination of chlorinated dibenzofurans associated with polychlorinated biphenyl fed to mallard (*Anas platyrhynchos*). Toxicol. Appl. Pharmacol., 37: 217–228.

Olafsdottir, K., K. Skirnisson, G. Gylfaottir and T. Johannesson, 1998. Seasonal fluctuations of organochlorine levels in the common eider (*Somateria mollissima*) in Iceland. Environ. Pollut., 103: 153–158.

Oxynos, K., J. Schmitzer and A. Ketrrup, 1993. Herring gull eggs as bioindicators for chlorinated hydrocarbons (Contribution to the German Federal Environmental Specimen Bank). Sci. Total Environ., 139–140: 387–398.

Pain, D.J., G. Burneleau, C. Bavoux and C. Wyatt, 1999. Levels of polychlorinated biphenyls, organochlorine pesticides, mercury and lead in relation to shell thickness in marsh harrier (*Circus aeruginosus*) eggs from Charente-Maritime, France. Environ. Pollut., 104: 61–68.

Parkinson, A. and S. Safe, 1987. Mammalian biologic and toxic effects of PCBs. In: S. Safe (Ed.), Polychlorinated biphenyls (PCBs): Mammalian and Environmental Toxicology (Environmental Toxin Series I), Springer, Berlin, pp. 49–75.

Parslow, J.L.F. and D.J. Jefferies, 1977. Gannets and toxic chemicals. Brit. Birds, 70: 366–372.

Peakall, D.B., 1974. DDE: its presence in peregrine eggs in 1948. Science (Wash. DC), 183: 673–674.

Peakall, D.B., 1990. Prospects for the peregrine falcon, *Falco peregrinus*, in the nineties. Can. Field Nat., 104: 168–173.

Peakall, D.B. and L.F. Kiff, 1988. DDE contamination in the peregrines and American kestrels and its effect on reproduction. In: T.J. Cade, J.H. Enderson, C.G. Thelander and C.M. White (Eds.), Peregrine Falcon Populations: Their Management and Recovery, Peregrine Fund, Boise, Idaho, pp. 337–350.

Ramesh, A., S. Tanabe, K. Kannan, A.N. Subramanian, P.L. Kumaran and R. Tatsukawa, 1992. Characteristic trend of persistent organochlorine contamination in wildlife from a tropical agricultural watershed, South India. Arch. Environ. Contam. Toxicol., 23: 26–36.

Ratcliffe, D.A., 1967. Decrease in eggshell weight in certain birds of prey. Nature (Lond.), 215: 208–210.

Sakamoto, K.Q., T. Kunisue, M. Watanabe, Y. Masuda, H. Iwata, S. Tanabe, F. Akahori, M. Ishizuka, A. Kazusaka and S. Fujita, 2002. Accumulation patterns of polychlorinated biphenyl congeners and organochlorine pesticides in Steller's sea eagles and white-tailed sea eagles, threatened species, in Hokkaido, Japan. Environ. Toxicol. Chem., 21: 842–847.

Scharenberg, W., 1991. Cormorants (*Phalacrocorax carbo sinenesis*) as bioindicators for polychlorinated biphenyls. Arch. Environ. Contam. Toxicol., 21: 536–540.

Schwarz, T.R. and D.L. Stalling, 1991. Chemometric comparison of poly-chlorinated residues and toxicologically active biphenyl congeners in the eggs of Forster's terns (*Sterna forsteri*). Arch. Environ. Contam. Toxicol., 20: 183–199.

Senthilkumar, K., K. Kannan, S. Tanabe and M. Prudente, 1998. Butyltin compounds in resident and migrant birds collected from Philippines. Fres. Environ. Bull., 7: 561–571.

Senthilkumar, K., K. Kannan, J.P. Giesy and S. Masunaga, 2002. Distribution and elimination of polychlorinated dibenzo p-dioxins, dibenzofurans, biphenyls and *p,p'*-DDE in tissues of bald eagles from the upper peninsula of Michigan. Environ. Sci. Technol., 36: 2789–2796.

Sethuraman, A. and AN. Subramanian, 2003. Organochlorine residues in the Avifauna of Tamil Nadu (Southeast coast of India). Chem. Ecol., 19 : 247-261.

Springer, A.M., W. Walker II, R.W. Risebrough, D. Benfield, D.H. Ellis, W.G. Mattox, D.P. Mindell and D.G. Roseneau, 1984. Origins of organo-chlorines accumulated by peregrine falcons, *Falco peregrinus*, breeding in Alaska and Greenland. Can. Field Nat., 98: 159–166.

Stickel, L.F., W.H. Stickell and R. Christensen, 1966. Distribution of DDT residues in brains and bodies of birds that died on dosage and in survivors. Science (Wash. DC), 151: 1549–1551.

Tanabe, S., 2002. Contamination and toxic effects of persistent endocrine disrupters in marine mammals and birds. Mar. Pollut. Bull., 45: 69–77.

Tanabe, S., A.N. Subramanian, H. Hidaka, H. and R. Tatsukawa, 1986. Transfer rates and patterns of PCB isomers and congeners and *p,p'*-DDE from mother to egg in Adelie penguin (*Pygoscelis adeliae*). Chemosphere, 15: 343–351.

Tanabe, S., S. Watanabe, H. Kan and R. Tatsukawa, 1988. Capacity and mode of PCB metabolism in small cetaceans. Mar. Mam. Sci., 4: 103–124.

Tanabe, S., K. Senthilkumar, K. Kannan, and A.N. Subramanian, 1998. Accumulation features of polychlorinated biphenyls and organochlorine pesticides in resident and migratory birds from South India. Arch. Environ. Contam. Toxicol., 34: 387–397.

Tanabe, S., M. Watanabe, T.B. Minh, T. Kunisue, S. Nakanishi, H. Ono and H. Tanaka, 2004. PCDDs, PCDFs and coplanar PCBs in albatross from the North Pacific and the Southern Ocean: levels, patterns and toxicological implications. Environ. Sci. Technol., 38: 403–413.

Thyen, S., P.H. Becker and H. Behmann, 2000. Organochlorine and mercury contamination of little terns (*Sterna albifrons*) breeding at western Baltic Sea. Environ. Pollut., 108: 225–238.

Tillitt, D.E., G.T. Ankley, D.A. Verbrugge, J.P. Giesey, J.P. Ludwig and T.J. Kubiak, 1991. H4IIE rat hepatoma cell bioassay-derived 2,3,7,8-tetrachlorodibenzo-p- dioxin equivalents in colonial fish-eating bird eggs from the Great Lakes. Arch. Environ. Contam. Toxicol., 21: 91–101.

Tillitt, D.E., G.T. Ankley, J.P. Giesey, J.P. Ludwig, H. Kuirta-Matsuba, D.V. Weseloh, P.S. Ross, A. Bishop, L. Seleo, K.L. Stromborg, L. Larson and T.J. Kubiak, 1992. Polychlorinated biphenyl residues and egg mortality in double-crested cormorants from the Great Lakes. Environ. Toxicol. Chem., 11: 1281–1288.

Vetter, W., U. Klobes and B. Luckas, 2001. Distribution and levels of eight toxaphene congeners in different tissues of marine mammals, birds and cod livers. Chemosphere, 43: 611–621.

Walker, C.H., C.G. Knight, J.K. Chipman and M.J.J. Ronis, 1984. Hepatic microsomal monooxygenases in sea birds. Mar. Environ. Res., 14: 416–419.

Wyk, E.V., H. Bouwman, H. van der Bank, G.H. Vendoorn and D. Hofmann, 2001. Persistent organochlorine pesticides in blood and tissue samples of vultures from different localities in South Africa. Comp. Biochem. Physiol., Part C, 129: 243–264.

Yamada, H., K. Takayanagi, M. Tateishi, H. Tagata and K. Ikeda, 1997. Organotin compounds and polychlorinated biphenyls of livers in squid collected from coastal waters and open oceans. Environ. Pollut., 2: 217–226.

Yamashita, N., S. Tanabe, J.P. Ludwig, H. Kurita, M.E. Ludwig and R. Tatsukawa, 1993. Embryonic abnormalities and organochlorine contamination in double-crested cormorants (*Phalacrocorax auritus*) and Caspian terns (*Hydroprogne caspia*) from the upper Great Lakes in 1988. Environ. Pollut., 79: 163–173.

Marine Mammals: Indicate Global Pollution by POPs and their Toxic Effects

The recent mass strandings and epizootics in marine mammals are correlated with high concentrations of POPs such as DDTs and PCBs in their bodies. Seas and oceans are final reservoirs of POPs. Being at the top of the marine food chain, marine mammals accumulate heavy loads of these contaminants in their lipid-rich tissues. Thus, they become suitable bioindicators of pollution by POPs.

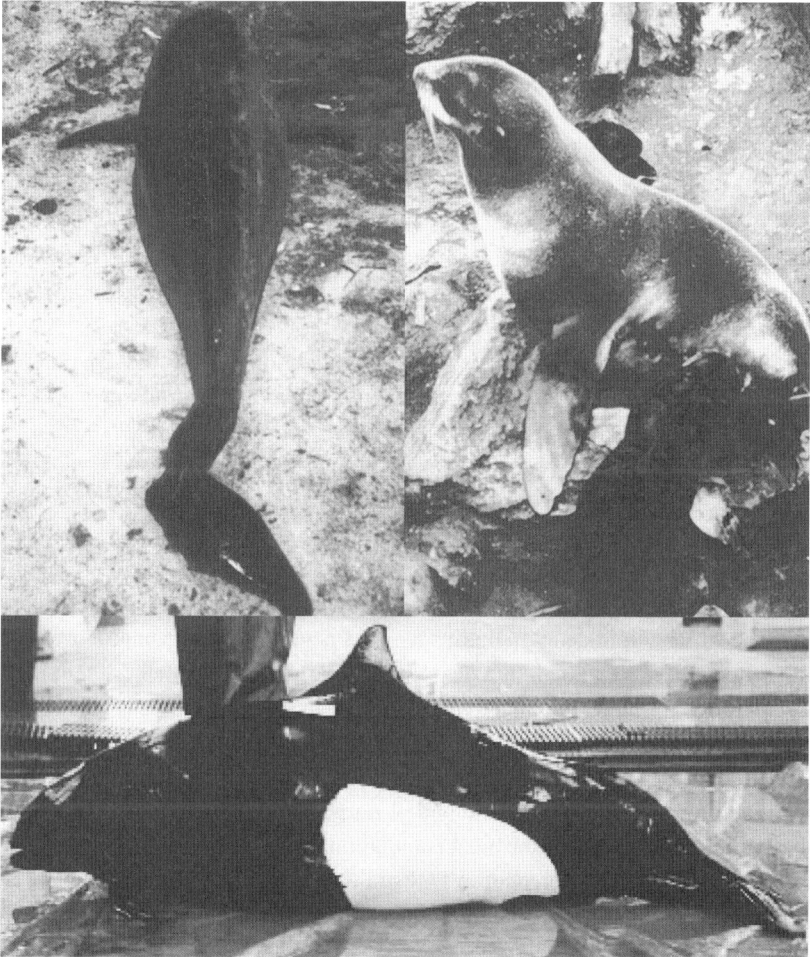

Chapter 6: Marine Mammals

Introduction

The global distribution of POPs and higher POP burdens in the tissues and organs of marine mammals due to their position at the top of the aquatic food chain and their relatively long lifespans are well established facts (Prudente et al., 1997; Wageman and Muir, 1984; Tanabe et al., 1984; Mossner and Ballschmiter, 1997; Vos et al., 2003). All the POPs that have been identified for immediate action by the Stockholm Convention are highly lipophilic, and the lipid reserves in the subcutaneous layers of marine mammals act as store houses for these chemicals (Tanabe et al., 1981, 1983).

Since the early 1930s, particularly following World War II, there has been large-scale production and widespread use of OCs in the form of pesticides, industrial chemicals, plasticizers, etc. There has been a sharp rise in the amounts of these chemicals entering the aquatic environment. Over the years, POPs have gradually infiltrated freshwater systems and have ultimately entered marine systems to become an integral part of the global marine environment. All the 12 POPs inevitably get accumulated in the tissues and organs of marine mammals. Long lifespans, migration, quantity and type(s) of food organisms, fat storage, relatively lower metabolic potential, etc. result in these organisms ingesting, accumulating and storing large of these chemicals. Marine mammals have a relatively long lifespan when compared with most of the other marine organisms; hence, the marine mammals have a long exposure time for these chemicals. The lipid depot in marine animals is very extensive in proportion to their body size, weight and the quantity of food intake; this makes them ideal repositories for large amounts of all lipophilic POPs.

The first ever report on OCs in marine mammals was published by Holden and Marsden (1967) who found considerable concentrations of chlorinated hydrocarbons in harbor seals and harbor porpoises. Following this, there have been numerous reports on the bioaccumulation, tissue distribution, metabolism, toxicity and gender differences in POPs in cetaceans and pinnipeds; hence, an enormous amount of information on this subject is currently available (Vos et al., 2003).

The availability of a large number of specimens during the 1970s and 80s due to the existence of commercial whaling in those years, the ease with which the POPs in the lipid stores of these animals could be analyzed and the advanced instruments and methodologies developed during that period led to innumerable publications on this subject. Moreover, the increasing numbers of mass mortalities of marine mammals (both cetaceans and pinnipeds) have been attributed to the larger xenobiotic loads in their bodies. Evidently, historical information on marine mammals reveals several mass mortalities that were widely separated over time and location (Lavigne and Schmitz, 1990). However, there has been a greater incidence of epizootics in the past 30 years. The recent seal, dolphin and porpoise mass mortalities occurred in a wide variety of ecosystems, and the animals in most of these mortalities contained the highest OC concentrations ever reported (Aguilar, 1991; Tanabe and Tatsukawa, 1991; Tanabe, 1994; Colborn and Smolen, 1996). This has led marine mammal scientists and environmentalists to concentrate their efforts on screening marine mammal for xenobiotics.

Colborn and Smolen (1996) listed 65 epidemiological abnormalities such as endocrine disruption, immunologic and reproductive dysfunction, tumours and population decline encountered in marine mammal populations since 1968. Simmonds (1991) stated that out of the 11 outbreaks of mass mortalities of marine mammals, 9 have occurred after the 1970s around the coastal waters of developed and industrialized countries, implicating toxic contaminants, including POPs. A report by Tanabe et al. (1994) critically evaluated the ecotoxicological impact of POPs on marine mammals.

Elevated POP concentration has been considered as one of the possible reasons for the reduction in immunity to canine distemper virus (CDV) in Caspian seals and Phocine distemper virus in harbor seals of Europe (Dietz et al., 1989). A possible reduction in the fecundity of the former seal population has also been attributed partly to POPs such as PCBs and DDTs. Marine mammals can thus be considered as important species, particularly for monitoring the long-term effects of pollution by POPs in the marine environment worldwide. They can also be used as global pollution indicators. Mossner and Ballschmiter (1997) are of the opinion that marine mammals can be considered as 'model systems' for studying low-dose, long-term effects of environmental pollution by persistent chemicals. Different species of these animals migrate to varying distances and in different routes; therefore, their pollutant loads integrate inputs from different region, even if local or regional pollution levels may not be explained, in many cases, by the pollutant concentrations in their bodies.

Among POPs, data are available on concentrations of major PCB congeners, DDTs, CHL and HCB in marine mammals. Toxaphene components and cyclodines (dieldrin and endrin) are less frequently measured. PCDDs and PCDFs have been detected in some marine mammal species, but limited information is available regarding their spatial trends (UNEP Chemicals, 2003). POPs such as DDTs, CHL, toxaphene, HCB, PCBs, PCDDs and PCDFs have been detected in marine mammal species worldwide (Oehme et al., 1988; Luckas et al., 1990; Iwata et al., 1993; Tanabe et al., 1994; Prudente et al., 1997; AMAP, 1998). These chemicals have been detected in almost all the species of cetaceans and pinnipeds collected even from remote areas such as the Arctic and Antarctic (Aono et al., 1997; Norstrom et al., 1998; Muir et al., 2000). Several reviews on the contamination and toxicity of POPs in marine mammals are available. In his book 'The Ecology of Whales and Dolphins', Gaskin (1982) discussed the occurrence and possible significance of environmental contaminants in cetaceans. Wagemann and Muir (1984) gathered the then available data regarding OCs in marine mammals in the northern waters and successfully evaluated the spatial variations and temporal trends in these chemicals in their bodies. Following this, Colborn and Smolen (1996) discussed the available background information, contaminant burdens, time order and specificity to effects of OCs with respect to the tissues and organs of marine mammals while reviewing the epidemiology of persistent OC contaminants in cetaceans.

Evaluating the available literature on the factors affecting the accumulation of POPs in marine mammals and the recent developments in the attempts to use marine mammals as bioindicators of pollution by POPs have raised some issues related to the use of these animals as bioindicators for routine monitoring of pollution by POPs in developing Asian countries. These include ethical issues related to sampling of marine mammals and variations associated with sampling. Nonetheless, the POP concentrations in the bodies of these animals can be successfully used for determining life-history parameters that may otherwise be difficult to determine. In addition, the loads of POPs in marine mammals can be used as bioindicators in specialized situations, a detailed discussion of which is given below.

Factors Affecting Accumulation

Tissue Distribution

All POPs are lipophilic in nature. As already stated, marine mammals have long lifespans and develop large lipid or fat reserves in their subcutaneous layers. Further, many of them occupy high trophic levels in the aquatic food chains, and most of them feed on lipid-rich food organisms. Blubber is the main repository for POPs. Tanabe et al. (1981) showed that the bodies of striped dolphins (*Stenella coeruleoalba*) are composed of, on average, 17.3% blubber, which contained the majority (over 90%) of the chlorinated hydrocarbon burden. Hence, they concluded that blubber can be used as the representative tissue to determine the dynamics and fate of chlorinated hydrocarbons in the bodies of marine mammals. Attempts were also made to quantify POP concentrations in various organs (e.g. liver) of marine mammals (Tanabe et al., 1984). At present, it is customary to use blubber tissue of marine mammals for monitoring the bioaccumulation of POPs.

Measurement of POPs in blubber may provide an easy and accurate method for evaluating the burdens of these chemicals in marine mammals. However, changes in the nutritional status of an individual animal may affect the concentrations of these compounds in lipid-rich tissues. Blubber thickness may be used as an indicator of the nutritional condition. Lipid weight and content are important parameters, particularly in emaciated individuals, because lipid mobilization in sick individuals may be associated with increased water content of blubber (Beck et al., 1993; Fadley, 1997), and variations in lipid types may affect the partitioning of POPs. Law (1994) stated that the collection of these biological data is important for toxicological interpretation of the results, particularly with respect to compounds that are lipophilic and susceptible to change with changes in the lipid status of the animal. Subramanian (1988) found that the PCB and DDE concentrations were higher in harbor porpoises with below-normal blubber thickness than in animals with normal complement of blubber. The patterns of distribution of PCB isomers and congeners also differed between normal and sick individuals. Such a phenomenon reiterates the fact that the data obtained from sick and stranded marine mammals, specimens with unknown cause of death and animals collected from events of mass mortality should be treated with caution, particularly while evaluating temporal and spatial variations in POPs.

Further, Kajiwara et al. (2001) observed a significant negative correlation between DDT and PCB concentrations and lipid content in specimens of diseased California sea lions (Fig. 6.1). The decrease in lipid content may be a result of fat mobilization due to a state of negative energy balance in diseased animals. Although OCs are expected to be mobilized along with lipids, the extent of mobilization of OCs in these animals was apparently less than that of the lipids. Aguilar (1985) and Beckmann et al. (1997) observed the same type of inverse relationship between blubber thickness or mass and OC concentrations in the blubber from diseased marine mammals.

Age Trend

Marine mammals generally have long lifespans (approximately 30 years) and accumulate POPs throughout their lives, starting from gestation and suckling stage. The general features of OC accumulation in marine mammals are fairly well understood. Generally, POP concentrations are similar in immature

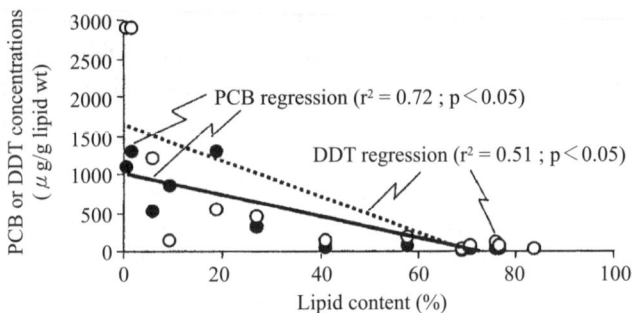

Fig. 6.1. Relationship between PCB or DDT concentrations and blubber lipid content in California sea lions (Source: Kajiwara et al., 2001).

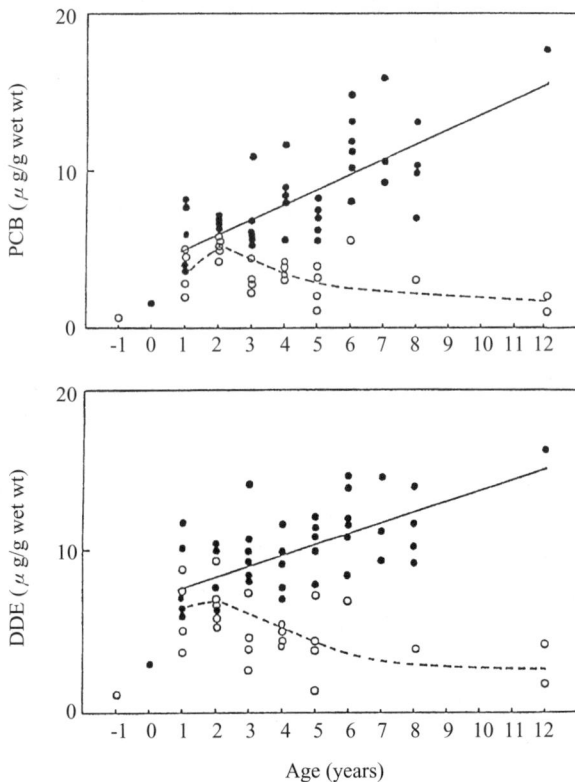

Fig. 6.2. Relationship between age and PCB and DDE concentrations in the blubber of dalli-type Dall's porpoises collected from northwestern North Pacific. (●): Males; (○): Females (Source: Subramanian et al., 1988).

males and females and increase with age until sexual maturity; after this, they continue to increase in males but either plateau or even slightly decrease in females (O'Shea and Tanabe, 1999). This behavior of POPs has been reported in several marine mammals such as harp seals (Addison et al., 1973), minke whales (Tanabe et al., 1986), short-finned pilot whales (Tanabe et al., 1987), Dall's porpoises (Subramanian et al., 1988) (Fig. 6.2), white whales (Stern et al., 1994), harbor porpoises (Westgate et al., 1997) and harbor and grey whales (Bernt et al., 1999). Such a long-term and continuous exposure to pollutants may render these animals unsuitable for use in monitoring the short-term changes in POP levels in the ambient environment. While attempting to utilize the POP data obtained from these animals for explaining geographical variations and/or temporal trends in the environment, care should be taken to consider sex, age and growth stages of the candidate species.

Gender Differences

Male-female differences in persistent lipophilic compounds are extremely prominent in marine mammals. Females of marine mammals excrete large quantities of persistent chemicals during the reproductive process (Subramanian, 1988). For example, a female grey seal may reduce 15%–30% of its OC load during the time it takes to raise and wean a single pup (Addison and Brodie, 1977, 1987). Citing the studies of Cockroft et al. (1989) and Addison and Stobo (1993), Beckmen et al. (1999) reported considerable transfer of OCs from mother to pup in various species of marine mammals. Similarly, large-scale transfer of OCs occurred from mother to calf in striped dolphins (Fukushima and Kawai, 1981). Further, Tanabe et al. (1981) found that in striped dolphins, more than 60% of the total body burden of OCs in the mother was transferred to the pup during a single lactation period. A female beluga whale in the St. Lawrence River estuary delivered 10 μg/ml of PCBs to her calf through her milk fat (Beland and Martineau, 1988).

Relocation of pollutants during pregnancy is not as important as that during lactation. In grey seals, the mother transferred only approximately 1% of her pollutant burden to her foetus (Donkin et al., 1981), while in weddell seal, it was 2% (Hidaka et al., 1983). On the other hand, Duinker and Hillebrand (1979) reported a significant transplacental transfer of OCs to the foetus in harbor porpoises (*Phocaena phocaena*), and Tanabe et al. (1982) assumed a transplacental transfer rate of less than 10% in striped dolphins.

In female Dall's porpoises, a significant decrease in PCB and DDE concentrations was observed from the age of 2 years, and this continued up to the active reproductive age of the species (Subramanian et al., 1988). Kajiwara et al. (2002) found higher OC concentrations in female Dall's porpoises without any corpora albicantia (no ovulation) than in those with corpora albicantia

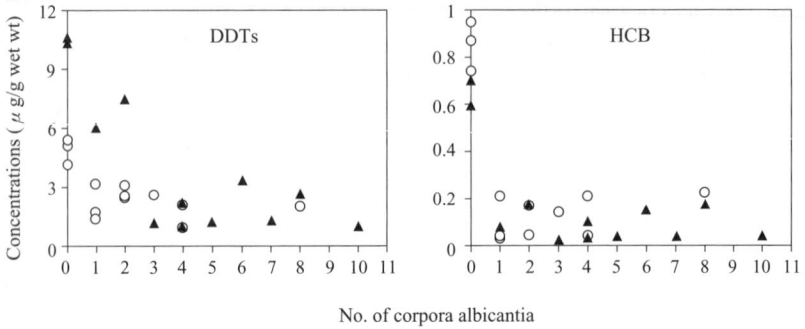

Fig. 6.3. Relationships between the number of corpora albicantia and the concentrations of DDTs and HCB in female Dall's porpoises. ○: truei-type; ▲: dalli-type (Source: Kajiwara et al., 2002).

(Fig. 6.3), indicating that females reduce OC residues through reproduction. In a population of striped dolphins affected by an epizootic, Borrell et al. (1997) found that the PCB and 2,3,7,8-TCDD concentrations as well as their TEQs in males were approximately double those in females. They observed the same trend in the concentrations of all individual PCB congeners.

When comparing the CRs of PCBs in adult females to adult males, Tanabe et al. (1986) found a higher mean value in minke whales (0.42) than in striped dolphins (0.24) (Tanabe et al., 1981) and Dall's porpoises (0.25) (Subramanian et al., 1986); this indicated different but large-scale reproductive transfers of OCs in female cetaceans.

The reproduction-dependent route of excretion of POPs in adults and the exposure of neonates to excessive POP concentrations may be critical factors that should be considered when these organisms are used as bioindicators of pollution by POPs. Such a remarkable gender difference in marine mammals may give rise to difficulties in interpreting the data on the basis of spatial and temporal scales; therefore, this factor should be considered while using marine mammals as bioindicators of pollution by POPs in coastal and oceanic environments.

Geographical Variation and Migration Pattern

In their recent report on PTS, UNEP Chemicals (2003) stated that marine mammals are usually migratory species; therefore, their pollutant loads will integrate PTS inputs from different regions. For example, Wagemann and Muir (1984) found that OC concentrations in marine mammals varied depending on their migration routes. While attempting to determine the migration routes of Dall's and True's porpoises, Subramanian et al. (1988) found different population structures in the northern North Pacific and adjacent waters. In a recent work, Loughlin et al. (2002) noted clear differences in PCB concentrations and a marginal difference in the concentration of DDTs in fur seal specimens collected from St. George Island and St. Paul Island of Alaska. Hobbs et al. (2003) found a geographical similarity in the concentrations of PCBs and OC pesticides, namely, DDTs, CHLs and HCHs, in the blubber of 155 minke whales (*Balaenoptera acutorostrata*) from seven regions in the North Atlantic as well as from the North Sea and Barents Sea; they attributed this finding to the high mobility of whales and their feeding in multiple areas within the northeastern Atlantic. In their review on the geographical and temporal variations in the levels of OC contamination in marine mammals, Aguilar et al. (2002) found a clear geographical variation; the greatest OC loads were found in marine mammals from the temperate fringe of the Northern Hemisphere, Mediterranean Sea and certain locations along the western coasts of North America.

Feeding Habits

High POP concentrations are found in marine mammals at the top of the food chain. While evaluating OCs in the western North Pacific ecosystem, Tanabe et al. (1983) reported that the concentrations of chlorinated hydrocarbons and the ratios of their concentrations in organisms to those in seawater (BCF) increased in higher ranking predators; striped dolphins showed very high BCF value of up to 10^7 for both PCBs and DDTs.

Generally, fish-eating species of marine mammals tend to have higher OC concentrations than those that feed on crustaceans (krill, plankton, etc.) by filter-feeding through baleen plates (Tanabe et al., 1984). For example, while analyzing the persistent pollutants—PCBs, DDTs, CHLs, HCHs and HCB—in the blubber and diet of minke whales collected from the Antarctic Ocean and North Pacific, Aono et al. (1997) found that the residue concentrations (except HCB) were lower in the Antarctic minke whales than in those from the North Pacific. Apart from the fact that the levels of these pollutants are generally lower in the Southern Hemisphere than in the Northern Hemisphere, the authors attributed the differences to the specific feeding habits of these two populations of minke whales, i.e. the minke whales in the Antarctic feed mainly on lower trophic organisms, primarily euphasiids, whereas the northern population feeds mainly on fishes. The authors also found that similar to the minke whale (predator), lower concentrations of OCs, except HCB, were found in the krill (prey) from the stomach contents of minke whales in the Antarctic than in those in the North Pacific; this clearly indicates the importance of food preferences of marine mammals in determining their body burdens of POPs.

Ruus et al. (1999) found that prey preferences influenced the patterns of accumulated pollutants in different species of pinnipeds. They found a high bioconcentration of DDT, but not of HCHs and HCB, in marine organisms from fishes (sandeel) to mammals (harbor seal). Apart from prey preferences in higher trophic organisms, the bioaccumulation mechanisms are affected by biochemical factors such as the metabolic capacity of the organisms; this explains the lower bioaccumulation of HCHs and HCB in harbor seals (Russ et al., 1999). Therefore, it appears that food preferences and metabolism should be considered while discussing the bioaccumulation of contaminants in marine mammals.

The accumulation of POPs in marine mammals differs according to their life stage due to changes in their feeding habits. For example, Addison and Stobo (1993) found that during the first year of life, grey seal pups accumulate virtually all OC residue burdens from their mothers' milk. They also found that metabolic degradation and excretion did not play any significant role in the deposition of OCs during this period of life. After weaning, the grey seals switch to a diet comprising organisms at a lower trophic level; hence, there was no great increase in the OC concentrations in proportion to the increase in body and blubber weights. In killer whales (*Orcinus orca*), the OC concentrations were higher in transient (migratory) individuals feeding primarily on other marine mammals (Hayteas and Duffield, 2000) than in resident individuals feeding mainly on fish (Jarman et al., 1996). Ross et al. (2000) found considerable PCB concentrations in three different populations of killer whales—the most contaminated cetaceans in the world—in the coastal waters of British Columbia; higher levels of contamination occurred in the transient individuals feeding on other marine mammals than in the resident whales feeding on fishes. Similarly, Muir et al. (1995) found higher OC concentrations in seal-eating walruses from Inukjuak than in Igloolik walruses and ringed seals feeding on pelagic invertebrates and bivalves.

Metabolism

Tanabe et al. (1988) reported that small cetaceans have a low metabolic capacity for a group of PCBs with adjacent non-chlorinated meta and para carbons in biphenyl rings. In particular, the PB-type enzymes are absent in small cetaceans, and the MC-type enzyme system is weak. Small cetaceans are capable of metabolizing some of the lower chlorinated biphenyls, and the authors believed that all small cetaceans might have the same capacity. Walker (1980) reported that hepatic microsomal monooxygenase activities are apparently lower in fish-eating birds than in rats. Similarly, Tanabe et al. (1988) found low enzyme activities in piscivorous small cetaceans, seals and minks (Fig. 6.4). Hence, a tendency of higher bioaccumulation of persistent OCs was noticed in small cetaceans. Further, Watanabe et al. (1989) found varying levels of monooxygenase activities in the fetus as well as the immature and mature individuals of short-finned pilot whales. The authors found no gender differences in CYP-linked monooxygenase activity in cetaceans.

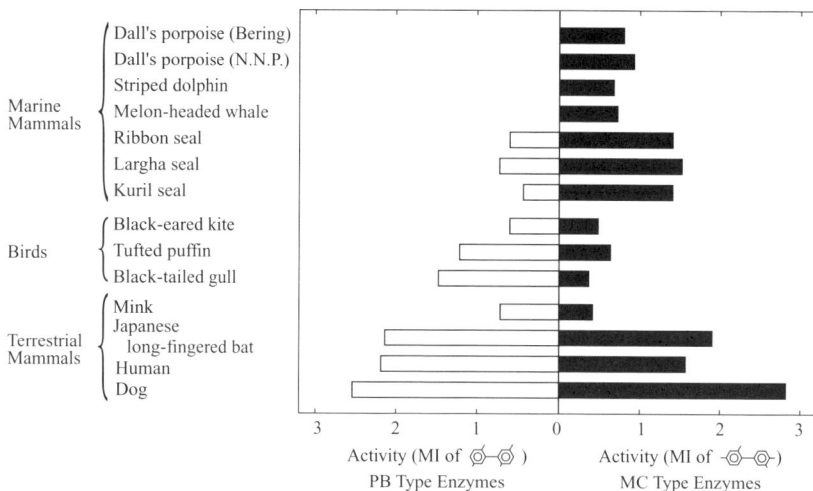

Fig. 6.4. Phenobarbital (PB)- and 3-methylcholanthrene (MC)-type enzyme activities in higher animals estimated by using the metabolic index (MI) of 2,2',5,5'- and 2,3',4,4'- tetrachlorobiphenyl isomers (Source: Tanabe et al., 1988).

After evaluating OC contaminants in the food chains in Jarfjord, Norway, Ruus et al. (1999) stated that although fish bioconcentrated OCs from water, harbor seals had the capability to metabolize and eliminate them. Shugart et al. (1990) demonstrated the induction of benzo(a)pyrene in the white whale population of St. Lawrence (a polluted area); however, benzo(a)pyrene was absent in the same species from the Arctic (low or less polluted areas). Correlation between PCB concentrations and mixed-function oxidase (MFO) induction has been demonstrated in cetaceans (Watanabe et al., 1989; White et al., 1994). Interestingly, Kannan et al. (1994) found a lower metabolic capacity for PCBs in the Ganges river dolphins (*Platanista gangetica*) than in marine and terrestrial mammals due to lower concentrations of CYP enzymes in their bodies.

Monitoring Studies

There are innumerable reports on the existence of POPs in marine mammals, particularly in lipid-rich tissues. In the past decade, the developed world has witnessed an enormous growth in the amount and quality of available literature on monitoring of POPs in marine animals due to the availability of large numbers of specimens obtained through commercial whaling, stranding, accidental catches and intentional scientific whaling as well as the development of accurate and chemical-specific analytical methods.

While evaluating contaminant concentrations and burdens of OC residues in harp seals from the Gulf of St. Lawrence and Hudson Strait, Beck et al. (1994) found some geographic variations in DDT and PCB burdens among female harp seals; a lower burden was detected in specimens from the Hudson Strait than in those from the St. Lawrence estuary. They could not compare the contaminant values in males because of the limited number of specimens that were analyzed. Further, they stated that any conclusion would be premature because of the limited number of samples and the concern over the reproductive transfer of POPs in females.

The use of chemicals to discriminate populations of marine mammals has been discussed in a review by Aguilar (1987). Subramanian et al. (1988) successfully used the same technique for determining the population structures of Dall's porpoises in the North Pacific and surrounding waters. Storr-Hansen and Spliid (1993) found a geographical variation in the PCB congener patterns in the blubber of harbor seals collected from three separate locations in Denmark. The authors attributed this to (i) differences in the biotransformation via enzymatic metabolism of PCB congeners in different populations, (ii) environmental factors such as different food species available at the locations and (iii) genetic adaptations in different seal populations.

Krahn et al. (1997) found considerable variations in the OC concentrations (PCBs, DDTs, CHLs and dieldrin) in four species of Alaskan seals (bearded seal, harbor seal, northern fur seal and ringed seal) collected for inclusion in the US National Biomonitoring Specimen Bank as well as for analysis as part of the contaminant monitoring component of the US National Marine Fisheries Service's Marine Mammal Health and Stranding Response Program. The authors explained the variations on the basis of feeding differences, migratory habits and proximity of harbor seals' habitat to urban coastal areas.

Muir et al. (2000) found geographical trends in PCBs, DDTs, HCHs and HCB in the ringed seals collected from the circumpolar areas of the Arctic. They found a general agreement of the HCH concentrations in seals with those in the ambient water. High PCB concentrations were found in the specimens collected near Russia. Nakata et al. (1998) also found noticeably higher concentrations of PCBs and DDTs in the ringed seals collected from the Kara Sea of the Russian Arctic than in those from the Canadian and Norwegian Arctic.

Kajiwara et al. (2002) found higher OC concentrations in specimens of Caspian seals (suffered from epizootics) collected in 2000 than in those collected in 1998, despite the fact that the status of OC contamination in the Caspian Sea environment may have improved during these years. Because of the differences in the nutritive status of specimens collected in different years, the authors could not suggest any temporal or spatial trends in these animals; however, they suggested that immunosuppression in these seals may have been caused by the elevated OC concentrations.

Tanabe et al. (2003) have successfully conducted temporal trend studies on OC contamination by using Baikal and Caspian seals as bioindicators. The concentrations of DDTs in Baikal and Caspian seals were considerably low in 1998 when compared with those in 1992, indicating a rapid decline in the concentrations of DDTs during that period. The authors also compiled the data on cetaceans collected from previous surveys and compared the temporal trends in both pinnipeds and cetaceans; they found that the clearance of OCs is slower in the open ocean cetaceans than in pinnipeds, which inhabit coastal and inland waters. These findings indicate the suitability of pinnipeds and cetaceans as bioindicators for measuring the temporal trends in POPs in the open oceans and coastal environments.

When compared with the available data on pinnipeds, information regarding cetaceans appears to be limited; this difference is evidently due to the ease in obtaining pinniped samples. Minh et al. (1999) analyzed blubber samples from two cetacean species, namely, Indo-Pacific hump-backed dolphins and finless porpoises, (found stranded along Hong Kong coastal waters) for PCBs, HCHs, DDTs and HCB. They detected very high OC concentrations in comparison with global levels and attributed this to the continuous environmental input of these compounds from the Far East region, including Hong Kong.

Prudente et al. (1997) found spatial variations in the concentrations of PCBs, CHLs, DDTs, HCHs and HCB in 11 species of adult male odontoceti collected from the western North Pacific, Indian Ocean and nearby seas (Fig. 6.5). The authors found relatively high concentrations of DDTs in the tropical species, which was attributed to the local current usage of these compounds and their limited mobility via long-range atmospheric transport. The HCH concentrations were higher in animals inhabiting the cold and temperate waters than in those inhabiting tropical waters; this result was reflective of the atmospheric transport of HCHs from the tropical source areas to northern sinks. A similar pattern was observed for PCBs and CHLs, which was attributed to the ongoing discharge of these compounds from Japan and other developed countries in the Northern Hemisphere.

High DDT concentrations were found in cetaceans from the Sea of Japan as well as the coastal waters of Hong Kong and India, indicating serious marine pollution in industrialized Asian nations and current usage of DDT in tropical regions. In general, compared with cetaceans from tropical regions, those inhabiting cold and temperate waters contained relatively higher concentrations of PCBs, HCHs, CHLs and HCB; this is reflective of the atmospheric transport of these compounds from the tropical sources to northern sinks.

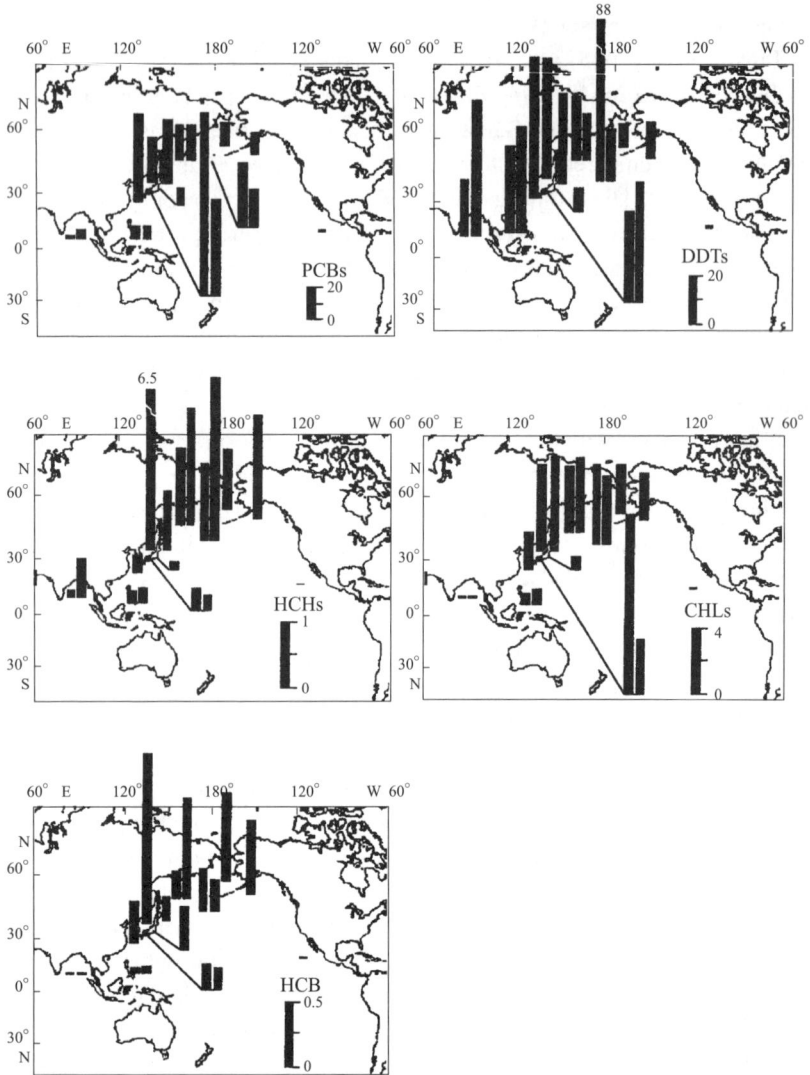

Fig. 6.5. Distribution of DDTs, PCBs, HCHs, CHLs and HCB residue concentrations (ìg/g wet wt) in odontoceti animals from various locations in the North Pacific, Indian Ocean and nearby seas (Source: Prudente et al., 1999).

Kannan et al. (1994) and Subramanian et al. (1999) detected high concentrations of OCs in blubber samples of the Ganges river dolphin (*Platanista gangetica*) comparable to the concentrations found in the three species of marine dolphins collected from Bay of Bengal (Tanabe et al., 1993; Karuppiah et al., 2005). They found a direct relationship between the concentrations of DDTs and PCBs in these dolphins and those in their prey species. They also detected an increase in OC levels from 1988 to 1992, indicating that increased pollutant loads were being added to the river Ganges.

Limitations

Although marine mammals are very good bioaccumulators of all the POPs (Tanabe et al., 1984) and are useful bioindicators for measuring long-term temporal trends and spatial variations, there are several factors that should be highlighted before recommending them for routine monitoring.

POP concentrations are always a few orders of magnitude higher in marine mammalian tissues than in the same organs of other marine animals. Marine mammals at the top of the food chain inadvertently biomagnify these chemicals from their prey organisms. However, this fact leads to the conclusion that the POP concentrations in marine mammalian tissues do not reflect the short-term changes in their ambient environment. As stated earlier, the purpose of this book is to recommend organisms for use as bioindicators, and based on the above-mentioned reasons, we suggest that marine mammals may not be suitable bioindicators for POP monitoring in the marine environment, except for specific purposes.

Obtaining suitable and adequate number of tissues for regular monitoring of POPs may be a difficult task. Tissue samples for pollution studies in cetaceans usually originate from three basic sources (i) strandings, (ii) direct and incidental catches and (iii) biopsies from free-ranging individuals (Reijnders et al., 1999). Stranded animals may have poor nutritional status and may contain abnormal pollutant concentrations either due to loss from the body or due to redistribution of pollutant loads from the blubber to other tissues. Regular monitoring of particular geographical region(s) cannot be accomplished using such specimens. Direct or incidental catches may be limited only to a small number of species and geographical regions. Fossi et al. (2003) recommended that instead of lethal approaches, non-lethal techniques such as obtaining skin biopsies should be used to evaluate residue concentrations and biomarker responses in species exposed to OCs with endocrine-disrupting capacity. Although, biopsy samples that are collected from a distance from free-ranging, apparently healthy specimens are devoid of the shortcomings of the samples collected from stranded and accidentally caught specimens, the former samples have limitations with respect to essential biological information such as age, sex and nutritional status, which cannot be obtained directly. Apart from these limitations, even if adequate numbers of samples can be obtained, the use of marine mammals for regular and short-term monitoring of POPs is impeded by the various facts and phenomena explained above.

All species of cetaceans and some pinnipeds are included in CITES (Convention on International Trade in Endangered Species of Wild Fauna and Flora) Appendices I and II, and their transport between countries is restricted through various regulations imposed by 160 parties. Pinnepeds inhabit only colder regions of the globe; hence, they may not be useful for monitoring the POP concentrations in developing countries, most of which are located in the tropical regions.

Apart from these facts, cetaceans and pinnipeds have several limitations such as their enormous body size, difficulties in dissection and obtaining sub-samples of various tissues and organs, wide range of migration, large male-female differences in residue concentrations, varying metabolic capacities among species and different natural histories.

Conclusions

Reijnders (1996) concluded that marine mammals are poor indicators of environmental contamination because of their low susceptibility to short-term changes in pollution. Marine mammalian tissues have thus far been used to determine the loads and possible adverse effects of toxic pollutants in their bodies; however, their use for continuous biomonitoring of short-term changes appears to be very difficult. Several parameters such as male-female differences, intra- and inter-species differences in metabolic capacities and differences in food preferences may make the interpretation of data tedious. At present, several other factors such as the availability of specimens, dissection and selection of appropriate tissues for analyzes of different POPs and transport of samples between laboratories in various countries appear complicated. At the same time, there is no doubt that marine mammals, both cetaceans and pinnipeds, are very good indicators for studying integrated temporal trends in POPs in the marine environments, particularly oceans.

References

Addison, R.F., S.R. Kerr and J. Dale, 1973. Variation of organochlorine residue levels with age in Gulf of St. Lawrence harp seals (*Pagophilus groenlandicus*). J. Fish. Res. Board Can., 30: 595–600.

Addison, R.F. and P.F. Brodie, 1977. Organochlorine residues in maternal blubber, milk and pup blubber from grey seals (*Halichoerus grypus*) from Sable Island, Nova Scotia. J. Fish. Res. Board Can., 34: 937–941.

Addison, R.F. and P.F. Brodie, 1987. Transfer of organochlorine residues from blubber through the circulatory system to milk in lactating grey seal *Halichoerus grypus*. Can. J. Fish. Aquat. Sci., 44: 782–786.

Addison, R.F. and W.T. Stobo, 1993. Organochlorine residue concentrations and burdens in grey seal (*Halichoerus grypus*) blubber during the first year of life. J. Zool. Lond. 230: 443–450.

Aguilar, A., 1985. Compartmentation and reliability of sampling procedures in organochlorine pollution surveys of cetaceans. Residue Rev., 95: 91–114.

Aguilar, A., 1987. Using organochlorine pollutants to discriminate marine mammal populations: a review and critique of the methods. Mar. Mamm. Sci., 3: 242–262.

Aguilar, A., 1991. Calving and early mortality in the western Mediterranean striped dolphin, *Stenella coeruleoalba*. Can. J. Zool., 69: 1408–1412.

Aguilar, A., A. Borrell and P.J.H. Reijnders, 2002. Geographical and temporal variation in levels of organochlorine contaminants in marine mammals. Mar. Environ. Res., 53: 425–452.

AMAP, 1998. AMAP Assessment Report: Arctic Pollution Issues. Arctic Monitoring and Assessment program (AMAP), Oslo, Norway, p. 859.

Aono, S., S. Tanabe, Y. Fujise, H. Kato and R. Tatsukawa, 1997. Persistent organochlorines in minke whale (*Balaenoptera acutorostrata*) and their prey species from the Antarctic and the North Pacific. Environ. Pollut., 98: 81–89.

Beck, C.G., T.G. Smith and M.O. Hammill, 1993. Evaluation of body condition in the north Atlantic harp seal (*Phoca groenlandica*). Can. J. Fish. Aquat. Sci., 50: 1372–1381.

Beck, G.G., T.G. Smith and R.F. Addison, 1994. Organochlorine residues in harp seals, *Phoca groenlandica*, from the Gulf of St. Lawrence and Hudson Strait: an evaluation of contaminant concentrations and burdens. Can. J. Zool., 72: 174–181.

Beckmann, K.B., L.J. Lowenstine, J. Newman, J. Hill, K. Hanni and J. Gerber, 1997. Clinical and pathological characterization of northern elephant seal skin disease. J. Wildl. Dis., 33: 438–449.

Beckmen, K.B., G.M. Ylitalo, R.G. Towell, M.M. Krahn, T.M. O'Hara and J.E. Blake, 1999. Factors affecting organochlorine contaminant concentrations in milk and blood of northern fur seal (*Callorhinus ursinus*) dams and pups from St. George Island, Alaska. Sci. Total Environ., 231: 183–200.

Beland, P. and D. Martineau, 1988. The beluga whale (*Delphinapterus leucas*) as integrator of St. Lawrence basin contamination history. In: Proceedings of the International Conference on Bio-indicators: exposure and effects, March 20–23, 1988, Oak Ridge National Laboratory, Knoxville, Tennessee.

Bernt, K.E., M.O. Hammill, M. Lebeuf and K.M. Kovacs, 1999. Levels and patterns of PCBs and OC pesticides in harbour and gray seals from the St. Lawrence estuary, Canada. Sci. Total Environ., 243–244: 243–262.

Borrell, A., A. Aguilar and T. Pastor, 1997. Organochlorine compound levels in striped dolphins from the western Mediterranean during the period 1987–1993. In: P.G.H. Evans (Ed.), European Research on Cetaceans, Vol. 10. pp. 281–285.

Cockroft, V.G., A.C. de Cock, D.A. Lordand and G.J.B. Ross, 1989. Organ-ochlorines in bottlenose dolphins (*Tursiops truncates*) from the east coast of South Africa. S. Afr. J. Mar. Sci., 8: 207–217.

Colborn, T. and M.J. Smolen, 1996. Epidemiological analysis of persistent organochlorine contaminants in cetaceans. Rev. Environ. Contam. Toxicol., 146: 91–172.

Dietz, R., M.P. Heidi-Jorgansen and T. Harkonen, 1989. Deaths of harbour seals in Europe. Ambio, 18: 258–264.

Donkin, P.S., V. Mann and E.I. Hamilton, 1981. Polychlorinated biphenyl, DDT and dieldrin residues in grey seal (*Halichoerus grypus*) males, females and mother-fetus pairs sampled at Farne Islands, England, during the breeding season. Sci. Total Environ., 19: 121–142.

Duinker, J. and M.T.J. Hillebrand, 1979. Mobilization of organochlorines from female lipid tissue and transplacental transfer to fetus in a harbour porpoise, *Phocoena phocoena*, in a contaminated area. Bull. Environ. Contam. Toxicol., 23: 728–732.

Fadley, B.S., 1997. Investigations of harbor seal health status and body condition in the Gulf of Alaska. Ph.D. Thesis, University of Alaska, Fairbanks, Alaska.

Fossi, M.C., L. Marsili, G. Neri, A. Natoli, E. Politi and S. Panigada, 2003. The use of a non-lethal tool for evaluating toxicological hazard of organochlorine contaminants in Mediterranean cetaceans: new data 10 years after the first paper published in MPB. Mar. Pollut. Bull., 46: 972–982.

Fukushima, M. and S. Kawai, 1981. Variation of organochlorine residue concentration and burden in striped dolphin, *Stenella coeruleoalba*, with growth. In: T. Fujiyama (Ed.), Studies on the Levels of Organochlorine

Compounds and Heavy Metals in the Marine Organisms, University of Ryukyus, Okinawa, pp. 97–114.

Gaskin, D.E., 1982. The Ecology of Whales and Dolphins. Heinemann, London and Exeter, New Hampshire, p. 459.

Hayteas, D.L. and D.A. Duffield, 2000. High levels of PCB and *p,p'*-DDE found in the blubber of killer whales (*Orcinus orca*). Mar. Pollut. Bull., 40: 558–561.

Hidaka, H., S. Tanabe and R. Tatsukawa, 1983. DDT compounds and PCB isomers and congeners in Weddell seals and their fate in the Antarctic marine ecosystem. Agric. Biol. Chem., 47: 2009–2017.

Hobbs, K.E., D.C.G. Muir, E.W. Born, R. Dietz, T. Haug, T. Metcalfe and N. Oien, 2003. Levels and patterns of persistent organochlorines in minke whale (*Balaenoptera acutorostrata*) stocks from the North Atlantic and European Arctic. Environ. Pollut., 121: 239–252.

Holden, A.V. and K. Marsden, 1967. Organochlorine pesticides in seals and porpoises. Nature, 216: 1274–1276.

Iwata, H., S. Tanabe, N. Sakai and R. Tatsukawa, 1993. Distribution of persistent organochlorines in the oceanic air and surface seawater and the role of ocean on their global transport and fate. Environ. Sci. Technol., 27: 1080–1098.

Jarman, W.M., R.J. Norstrom, D.C.G. Muir, B. Rosenberg, M. Simon and R.W. Baird, 1996. Levels of organochlorine compounds, including PCDDs and PCDFs in the blubber of cetaceans from the west coast of North America. Mar. Pollut. Bull., 32: 426–436.

Kajiwara, N., K. Kannan, M. Muraoka, M. Watanabe, S. Takahashi, F. Gulland, H. Olsen, A.L. Blankenship, P.D. Jones, S. Tanabe and J.P. Giesy, 2001. Organochlorine pesticides, polychlorinated biphenyls, and butyltin compounds in blubber and livers of stranded California sea lions, elephant seals, and harbor seals from coastal California, USA. Arch. Environ. Contam. Toxicol., 41: 90–99.

Kajiwara, N., S. Niimi, M. Watanabe, Y. Ito, S. Takahashi, S. Tanabe, L.S. Khuraskin and N. Miyazaki, 2002. Organochlorine and organotin compounds in Caspian seals (*Phoca caspica*) collected during an unusual mortality event in the Caspian Sea in 2000. Environ. Pollut., 117: 391–402.

Kannan, K., S. Tanabe, R. Tatsukawa and R.K. Sinha, 1994. Biodegradation capacity and residue pattern of organochlorines in Ganges River dolphins from India. Toxicol. Environ. Chem., 42: 249–261.

Karuppiah, S., A. Subramanian and J.P. Obbard. 2005. Organochlorine residues in odontocete species from the southeast coast of India. Chemosphere, 60: 891-897.

Krahn, M.M., P.R. Becker, K.L. Tribury and J.E. Stein, 1997. Organochlorine contaminants in blubber of four seal species: integrating biomonitoring and specimen banking. Chemosphere, 34: 2109–2121.

Lavigne, D.M. and O.J. Schmitz, 1990. Global warming and increasing population densities: a prescription for seal plagues. Mar. Pollut. Bull., 21: 280–284.

Law, R.J., 1994. Collaborative UK marine mammals project: summary of data produced 1988–1992. Fish. Res. Tech. Rep. 97, MAFF, Lowesttoft, UK, p. 42.

Loughlin, T.R., M.A. Castellini and G. Ylitalo, 2002. Spatial aspects of organochlorine contamination in northern fur seals. Mar. Pollut. Bull., 44: 1024–1034.

Luckas, B., W. Vetter, P. Fischer, G. Heidemann and J. Plotz, 1990. Characteristic chlorinated hydrocarbon patterns in the blubber of seals from different marine regions. Chemosphere, 21: 13–19.

Minh, T.B., M. Watanabe, H. Nakata, S. Tanabe and T.A. Jefferson, 1999. Contamination by persistent organochlorines in small cetaceans from Hong Kong coastal waters. Mar. Pollut. Bull., 39: 383–392.

Mossner, S. and K. Ballschmiter, 1997. Marine mammals as global pollution indicators for organochlorines. Chemosphere, 34: 1285–1296.

Muir, D.C.G., M.D. Segstro, K.A. Hoson, C.A. Ford, R.E.A. Stewart and S. Olpinski, 1995. Can seal eating explain elevated levels of PCBs and organochlorine pesticides in walrus blubber from eastern Hudson Bay (Canada)? Environ. Pollut., 90: 335–348.

Muir, D.C.G., F. Riget, M. Cleeman, J. Skaare, L. Kleivane, H. Nakata, R. Dietz, T. Severinsen and S. Tanabe. 2000. Circumpolar trends of PCBs and organochlorine pesticides in the Arctic marine environment inferred from levels in ringed seals. Environ. Sci. Technol., 34: 2431–2438.

Nakata, H., K. Kannan, L. Jing, N. Thomas, S. Tanabe and J.P. Giesy, 1998. Accumulation pattern of organochlorine pesticides and polychlorinated biphenyls in southern sea otters (*Enhydra lutris nereis*) found stranded along coastal California, USA. Environ. Pollut., 103: 45–53.

Norstrom, R.J., S.E. Belikov, E.W. Born, G.W. Garner, B. Malone, S. Olpinski, S. Ramsay, S. Schliebe, I. Stirling, M.S. Stishov, M.K. Taylor and O. Wiig, 1998. Chlorinated hydrocarbon contaminants in polar bears from eastern Russia, North America, Greenland and Svalbard: biomonitoring of Arctic pollution. Arch. Environ. Contam. Toxicol., 35: 354–367.

Oehme, M., P. Furst, C. Kruger, H.A. Meemken and W. Grobel, 1988. Presence of polychlorinated dibenzo-*p*-dioxins, dibenzofurans and pesticides in Arctic seal from Spitzbergen. Chemosphere, 17: 1291–1300.

O'Shea, T.J. and S. Tanabe, 1999. Persistent ocean contaminants and marine mammals: a retrospective overview. In: T.J. O'Shea et al. (Eds.), Proceedings of the Marine Mammal Commission Workshop, Marine Mammals and Persistent Ocean Contaminants, pp. 87–92.

Prudente, M., S. Tanabe, M. Watanabe, A.N. Subramanian, N. Miyazaki, P. Suarez, P. and R. Tatsukawa, 1997. Organochlorine contamination in some odontoceti species from the North Pacific and Indian Ocean. Mar. Environ. Res., 44: 415–427.

Reijnders, P.J.H., 1996. Organohalogen and heavy lethal contamination in cetaceans: observed effects, potential impact and future prospects. In: M.P. Simmonds and J.D. Hutchinson (Eds.), The Conservation of Whales and Dolphins. John Wiley & Sons Ltd., Chichester, UK, pp. 205–217.

Reijnders, P.J.H., G.P. Donovan, A. Aguilar and A. Bjorge, 1999. (Eds.). Report of the workshop on chemical pollution and cetaceans. J. Cetacean Res. Manage., Spec. Issue, p. 53.

Ross, P.S., G.M. Ellis, M.G. Ikonomou, L.G. Barrett-Lennard and R.F. Addison, 2000. High PCB concentrations in free-ranging Pacific killer whales, *Orcinus orca*: effects of age, sex and dietary preference. Mar. Pollut. Bull., 40: 504–515.

Ruus, A., K.I. Ugland, O. Espeland and J.U. Skaare, 1999. Organochlorine contaminants in a local marine food chain from Jarfjord, northern Norway. Mar. Environ. Res., 48: 131–146.

Shugart, L.R., D. Martineau and P. Beland, 1990. Detection and quantitation of benzo(a)pyrene adducts in brain and liver tissues of beluga (*Delphinapterus leucas*) from the St. Lawrence and Mackenzie estuaries. Press Universite du Quebec, pp. 219–223.

Simmonds, M., 1991. Marine mammal epizootics worldwide. In: X. Pastor and M. Simmonds (Eds.), Proceedings of the Mediterranean Striped Dolphin Mortality International Workshop, Greenpeace International Mediterranean Sea Project, Madrid, Spain, pp. 9–19.

Stern., G.A., D.C.G. Muir, M.D. Segstro, R. Dietz and M.P. Heide-Jorgensen, 1994. PCBs and other organochlorine contaminants in white whales (*Delphinapterus leucas*) from West Greenland: variations with age and sex. Bioscience, 39: 245–258.

Storr-Hansen, E. and H. Spliid, 1993. Coplanar polychlorinated biphenyl congener levels and patterns and the identification of separate populations of harbor seals (*Phoca vitulina*) in Denmark. Arch. Environ. Contam. Toxicol., 24: 44–58.

Subramanian, A.N., 1988. Persistent organochlorines as chemical tracers in determining the biological and ecological parameters of Dall's porpoises, *Phocoenoides dalli* Ph.D. Thesis, Ehime University, Japan, p. 137.

Subramanian, A.N., S. Tanabe, Y. Fujise and R. Tatsukawa, 1986. Organochlorine residues in Dall's and True's porpoises collected from northwestern Pacific and adjacent waters. Mem. Natl. Inst. Polar Res. 44: 167–173.

Subramanian, A.N., S. Tanabe and R. Tatsukawa, 1988. Chemical approach to determine some biological and physiological aspects of Dall's porpoises using organochlorines as tracers. Res. Org. Geochem., 6: 51–54.

Subramanian, AN., R.S. Lal Mohan, V.M.Karunagaran and R. Babu Rajendran, 1999. Concentrations of HCHs and DDTs in the tissues of River dolphins *Platanista gangetica.* Chem. Ecol., 16 : 143–150.

Tanabe, S., 1994. Fate of persistent organochlorines in the marine environment, Contaminants in the Environment. R. Renzoni, N. Mattei, L. Lali and M. C. Fossi (Eds.), CRC Press, Boca Raton, Florida, pp. 19–28.

Tanabe, S. and R. Tatsukawa, 1991. Persistent organochlorines in marine mammals. Organic Contaminants in the Environment, K. C. Jones (Ed.), Elsevier Applied Science, Cambridge, Great Britain, pp. 275–289.

Tanabe, S., R. Tatsukawa, H. Tanaka, K. Maruyama, N. Miyazaki and T. Fujiyama, 1981. Distribution and total burdens of chlorinated hydro-carbons in bodies of striped dolphins (*Stenella coeruleoalba*). Agric. Biol. Chem., 45: 2569–2578.

Tanabe, S., R. Tatsukawa, K. Maruyama and N. Miyazaki, 1982. Transplacental transfer of PCBs and chlorinated hydrocarbon pesticides from the pregnant striped dolphin (*Stenella coeruleoalba*) to her fetus. Agric. Biol. Chem., 46: 1249–1254.

Tanabe, S., T. Mori, R. Tatsukawa and N. Miyazaki, 1983. Global pollution of marine mammals by PCBs, DDTs and HCHs (BHCs). Chemosphere, 12: 1269–1275.

Tanabe, S., T. Mori and R. Tatsukawa, 1984. Bioaccumulation of DDTs and PCBs in the southern minke whale (*Balaenoptera acutrostrata*). Mem. Natl. Inst. Polar Res., 32: 140–150.

Tanabe, S., S. Miura and R. Tatsukawa, 1986. Variations of organochlorine residues with age and sex in Antarctic minke whale. Mem. Natl. Inst. Polar Res., 44: 174–181.

Tanabe, S., B.G. Loganathan, A.N. Subramanian and R. Tatsukawa, 1987. Organochlorine residues in short-finned pilot whale: possible use as tracers of biological parameters. Mar. Pollut. Bull., 18: 561–563.

Tanabe, S., S. Watanabe, H. Kan and R. Tatsukawa, 1988. Capacity and mode of PCB metabolism in small cetaceans. Mar. Mamm. Sci., 4: 103–124.

Tanabe, S., AN. Subramanian, A. Ramesh, PL. Kumaran, N. Miyazaki and R. Tatsukawa, 1993. Persistent organochlorine residues in dolphins from the Bay of Bengal, South India. Mar. Environ. Res., 26 : 311-316.

Tanabe, S., H. Iwata and R. Tatsukawa, 1994. Global contamination by persistent organochlorines and their ecotoxicological impact on marine mammals. Sci. Total Environ., 154: 163–177.

Tanabe, S., S. Niimi, T.B. Minh, N. Miyazaki and A.E. Petrov, 2003. Temporal trends of persistent organochlorines in Russia: a case study of Baikal and Caspian seal. Arch. Environ. Contam. Toxicol., 44: 533–545.

UNEP Chemicals, 2003. Regionally Based Assessment of Persistent Toxic Substances. Global Report 2003, p. 207, http: //www.chem.unep.ch/pts.

Vos, J.G., G.D. Bossart, M. Fournier and T.J. O'Shea, 2003. New perspectives: Toxicology and the Environment. Toxicology of Marine Mammals. Taylor & Francis, London and New York, p. 643.

Wageman, R. and D.C.G. Muir, 1984. Concentrations of heavy metals and organochlorines in marine mammals of northern waters: an overview and evaluation. Can. Tech. Rep. Fish. Aquat. Sci., 1279: 1–97.

Watanabe, S., T. Shimada, S. Nakamura, N. Nishiyama, N. Yamashita, S. Tanabe and R. Tatsukawa, 1989. Specific profile of liver microsomal cytochrome P-450 in dolphin and whales. Mar. Environ. Res., 27: 51–65.

Walker, C.H., 1980. Species variations in some hepatic microsomal enzymes that metabolize xenobiotics. Prog. Drug Metab., 5: 113–164.

Westgate, A.J., D.C.G. Muir, D.E. Gaskin and M.C.S. Kingsley, 1997. Concentrations and patterns of organochlorine contaminants in the blubber of harbour porpoises, *Phocoena phocoena*, from the coast of Newfoundland, the Gulf of St. Lawrence and the Bay of Fundy/Gulf of Maine. Environ. Pollut., 95: 105–119.

White, R.D., M.E. Hahn, W.L. Lockhart and J.J. Stegman, 1994. Catalytic and immunological characterization of hepatic microsomal cytochromes P450 in beluga whale (*Delphinapterus leucas*). Toxicol. Appl. Pharmacol., 126: 45–57.

Humans: Bioindicators Providing Most Relevant Data on Pollution

Humans are exposed to POPs through a variety of routes—air, water and food. Despite some social, ethical and legal impediments, human tissue samples collected in a non-destructive manner are the best samples for measuring spatial and temporal variations in POPs. The best advantage of using human tissue samples is that background data such as age, reproduction and possible routes of exposure can be accurately obtained from the subject(s).

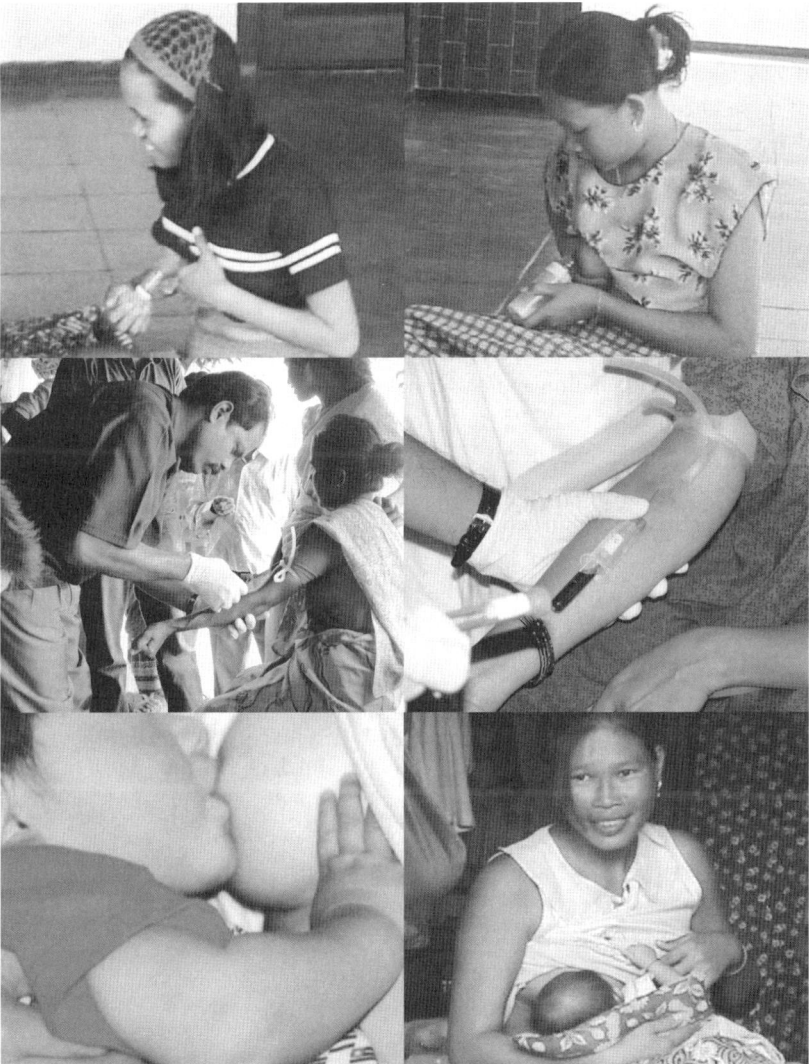

Chapter 7: Humans

Introduction

The presence of POPs in humans is viewed with concern because of their potential for teratogenic, carcinogenic, hormonal, neurological and immunological effects (Nicholson and Landrigan, 1994). POPs are persistent chemicals; therefore, even after imposing restrictions or at least reductions on their emissions, it may take a long time before a decline in human exposure and body burdens of these chemicals can be achieved. Considerable amount of human toxicity data on POPs derived from regular chronic exposure and acute accidental and occupational exposures are available (Loffler and van Bavel, 2000; Foster et al., 2002). There are various pathways through which humans are exposed to these pollutants. Besides dietary intake, which is considered as the most important source of pollutants, inhalation and dermal absorption are also of significance (Loffler and van Bavel, 2000). Marquardt and Schafer (1994) suggested that approximately 90% of the total intake occurs through diet, and the remaining 10% is through inhalation and/or dermal exposure.

Human exposure to POPs starts from the prenatal period in the mother's womb and continues until death. Humans may be exposed to several compounds simultaneously; these compounds may then get accumulated, become inactivated or may be eliminated (Schaefer et al., 2000). In spite of the innumerable reports on the occurrence of POPs in human tissues and organs, the mechanism of uptake and accumulation and the mode of action in humans remain to be elucidated.

Different human tissues have been used to measure the levels of exposure. The most commonly used tissues comprise adipose tissue, blood (serum) and human milk. Apart from these, a few reports describe the quantification in hair, liver, kidney, semen, muscle, meconium and endometrium. Most of these studies have been conducted to gain knowledge on the concentrations and toxic effects of these xenobiotics, but the data obtained revealed many geographic and temporal variations depending on the location and time of collection. Obtaining human tissues for regular monitoring of POPs in developing countries may be difficult. However, briefly reviewing the routes of exposure, patterns of contamination and their distribution in different human tissues will certainly be useful for interpreting the data obtained through analyzing other suitable bioindicator organisms. When available, human samples may also be used as bioindicators of POPs.

Factors Affecting Accumulation

Exposure

Diet is the main route of human exposure to POPs. The exposure starts from the prenatal period in the mother's womb. Foster et al. (2002) obtained amniotic fluid samples from 175 women undergoing routine amniocentesis between 14 and 21 weeks of gestation and analyzed the samples for several endocrine-disrupting compounds. They detected DDE in only 25% of the samples; HCH and PCBs were detected in some samples. However, several other authors have detected considerable levels of such estrogenic OCs in human amniotic fluid samples from countries such as India and Yugoslavia (Saxena et al., 1980; Siddiqui et al., 1981; Bazulic et al., 1984).

Human exposure to OCs occurring through breast milk is comparatively higher when compared with prenatal exposure. Recently, several workers have reported the exposure of infants to POPs through their first food intake. For example, Kunisue et al. (2004) reported high concentrations of DDTs, PCBs, HCHs, HCB and CHLs in all the 36 milk samples collected from the urban population of Phnom Penh city, Cambodia, during 1999–2000. They also found a considerable decrease in the concentrations of all these OCs in the breast milk with an increase in the number of children; this implies that the first infant may be exposed to higher OC concentrations through breast milk. There were several reports on the exposure of infants from various parts of the world, including Russia (Polder et al., 2003), Indonesia (Burke et al., 2003), Spain (Campoy et al., 2001), Jordan (Nasir et al., 1998), Thailand (Stuetz et al., 2001) and Zimbabwe (Chikuni et al., 1997).

After weaning, children are exposed to OCs through several types of diets. Humans are omnivores; hence, determining the route(s) of exposure without preplanned exposure experiments may be very difficult. For example, by analyzing 39 food items collected from China, Nakata et al. (2002) found that marine foods are the main source of DDT exposure in humans; vegetables and cereals, HCHs and vegetables and meat, HCB. As early as in 1990, Tanabe et al. (1990) found higher concentrations of HCHs in the breast milk of vegetarian women than of non-vegetarian women in India; this was later endorsed by the findings of Kannan et al. (1997) and Nakata et al. (2002).

Fish have been reported to be the prime sources of human exposure to PCBs in many developed countries (Noren, 1983). Smith and Gangolli (2002) stated that populations subsisting largely on fish and marine products as well as those that consume fish oils are at a higher risk of OC contamination. From the data collected from Japan, Yoshida et al. (2000) estimated that one-half of the daily intake of dioxins by the general population occurred through fish. In contrast, Tanabe et al. (1990) could not find high PCB concentrations in milk samples obtained from fish-eaters and fisherwomen of southern India. Kannan et al. (1992) found that fish and dairy products were the prime sources of OC exposure in India; the OC concentrations found in these food items were close to or exceeded the tolerance limits fixed by the FAO/WHO. Elevated concentrations of PCBs and CHLs were detected in Australian sea foods (Kannan et al., 1994). In Chinese food items, very high concentrations of HCHs and DDTs were detected in rice, fish and eggs (Li et al., 1998; Nakata et al., 2002). Kashyap et al. (1994) suggested that the use of DDT and HCH for the control of malaria in India leads to the contamination of cow's milk, food grains and vegetables, thereby leading to higher OC concentrations in the milk samples of vegetarian women (Tanabe et al., 1990).

Over a period of 12 years from 1986 to 1997, Schade and Heinzow (1998) analyzed 3500 human breast milk samples collected from women living in northern Germany. Women who consumed a balanced diet for at least 3 years were found to have lower HCB and β-HCH levels. Women who ate more than 100 g of fish or more than 700 g of meat per week were more likely to have higher PCB and β-HCH concentrations or HCB concentrations, respectively.

It may be difficult to determine the particular route(s) of exposure of humans to any of the POPs. However, human samples are superior to animal samples because the feeding history and other biological information of human subjects can be obtained with considerable accuracy.

Age

As in the case of all animals with long lifespans, age- and gender-dependent accumulation of OCs may normally be expected in humans. However, variations in the degree of exposure due to different dietary habits and different excretory routes and metabolic capacities for different POPs have resulted in conflicting reports on the age-dependent OC accumulation in humans.

Minh et al. (2001) analyzed human adipose tissue, liver and bile samples from Japan for the concentrations of PCBs, DDTs and HCB as well as HCHs, *tris*(4-chlorophenyl)methane (TCPMe) and *tris*(4-chlorophenyl)methanol (TCPMOH). They observed that, in general, the concentrations of all the compounds increased with age in both females and males (Fig. 7.1); higher concentrations of DDTs, HCHs and CHLs were observed in older persons than in the young.

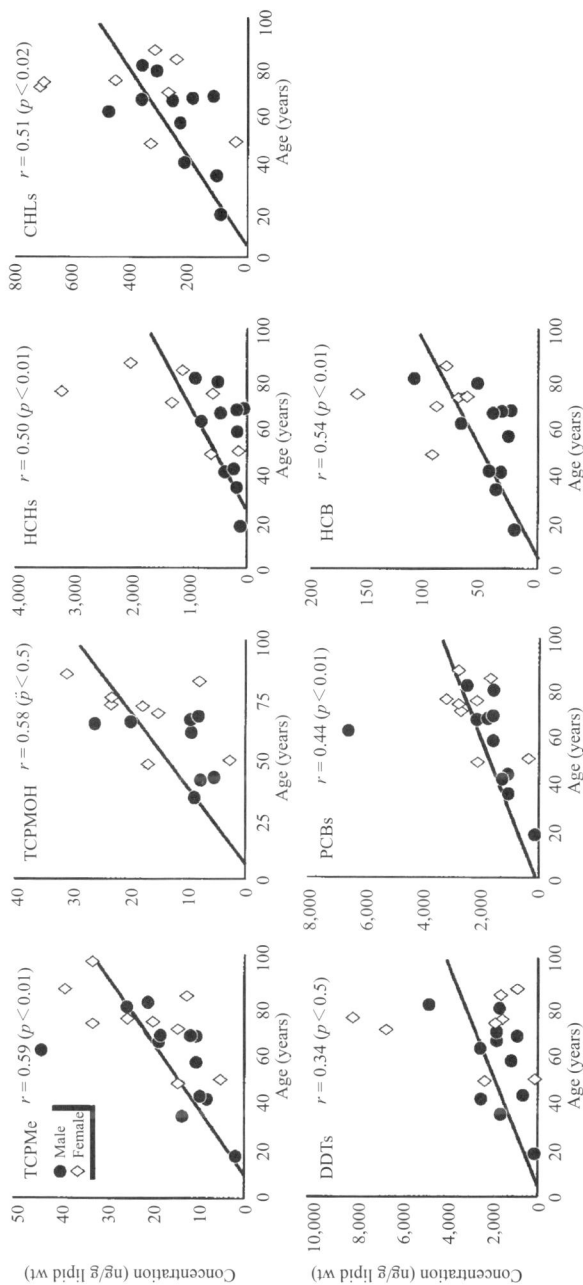

Fig. 7.1. Accumulation pattern of OCs with respect to age in Japanese human adipose tissue samples (Source: Minh et al., 2001).

Smeds and Saukko (2001) quantified endocrine-disrupting pesticides in human adipose tissue samples from Finland and found significant age-related accumulation of DDE in females and HCB in males. Covaci et al. (2002) found higher PCB concentrations (504 ng/g lipid weight) in 20 adipose tissue samples collected from Belgian males (mean age, 47.2 years) than those detected by Pauwels et al. (2000) in 46 adipose tissue samples (334 ng/g lipid weight) collected from Belgian males (mean age, 31.9 years); this confirmed that PCB burden increases with age. Interestingly, the authors could not find any age-related accumulation of PBDEs in their samples, while PCBs and DDTs showed higher correlations (P < 0.05) with age. On the other hand, Kang et al. (1997) analyzed human adipose tissue samples from Korea and found no significant correlation between age and residue concentrations of DDTs, PCBs, PCDDs and PCDFs, whereas HCHs showed a significant age-dependent increase.

Several factors such as reproductive transfer, metabolism and exposure may influence the accumulation of POPs in human tissues, which may in turn affect the age-dependent concentration of POPs in the whole body. All these factors should be considered while planning a sampling scheme for collecting human samples for POPs monitoring.

Gender Differences

Humans are exposed to POPs in the environment through contamination of food, air and water. Once these chemicals enter the body, they may undergo extensive biotransformation. Some chemicals are rapidly metabolized and excreted in urine and/or feces; others are slowly metabolized and extensively stored in tissues/organs. Apart from these routes of exposure, considerable quantities of these lipophilic chemicals may be excreted through reproductive pathways in females, leading to significant gender differences in the contaminant loads, as is observed in other mammalian species such as marine mammals (Subramanian et al., 1986; Tanabe, 2000).

Kunisue et al. (2004) studied the occurrence of OCs in human breast milk samples collected from Cambodia and observed that the OC concentrations decreased only slightly with the age of the women; however, they found that the OC concentrations tended to decrease with increase in the number of children, suggesting a possible route of excretion of lipophilic OCs in females. The authors also suggested that 50%–60% of the lipids in human breast milk are obtained from adipose tissue and hepatic lipids, 30% from ingested dietary lipids and 10%–20% from *de novo* mammary synthesis; therefore, mobilization of lipophilic OCs occurs from lipid deposits in the body to breast milk.

While analyzing adipose tissue samples of Jordanian males and females, Alawi et al. (1999) found higher concentrations of HCB, HCHs and DDTs in male subjects than in female subjects; this was attributed to the higher OC exposure in males than in females. However, the authors did not collect any information on the reproductive history of the subjects; hence, lactational transfer may also have been the main reason for such a difference.

In Portugal, Cruz et al. (2003) observed higher concentrations of HCHs and DDTs in the blood serum of males than in that of females from one urban and two rural populations; these differences were also observed in Belgium, Egypt and USA (DeVoto et al., 1998; Hanaoka et al., 2002; Manirakiza et al., 2002). The authors attributed the differences to lactational transfer as well as other gender differences in storage and metabolism of these compounds, as was suggested by Ahlborg et al. (1995).

In contrast, Smeds and Saukko (2001) found higher total pesticide concentrations in females (1910 ng/g lipid weight) than in males (478 ng/g lipid weight) from Finland. However, they did not observe any statistically significant differences in individual compounds between sexes; this is in accordance with the results of human adipose tissue samples from Italy (Gallelli et al., 1995).

Minh et al. (2001) observed no significant male-female difference in the concentrations of DDTs, PCBs, HCB, CHLs, TCPMe and TCPMOH in human adipose tissue samples from Japan. They stated that unlike in the case of aquatic mammals where large gender differences were observed (e.g. Subramanian et al., 1986; Nakata et al., 1995, 1998), the small gender differences observed in humans may be attributed to a shorter lactation period, lower lipid content of milk and smaller number of child births. Although male-female differences are indicated in many reports and could be more logically explained by the authors, there are contradicting reports that should also be taken into account while interpreting the data on POP monitoring using human tissues.

Metabolism and Excretion

The main effects of POPs in animals and humans are endocrine alterations, immunotoxicity and carcinogenicity. OCs may be absorbed through the gut, respiratory system and skin. Although OCs are initially concentrated in the blood, liver and muscle, the adipose tissue and skin are the long-term storage sites. Some of these compounds are metabolized by the liver, and the primary reaction is hydroxylation by several CYPs, followed by conjugation with glucuronic acid (Letz et al., 1990).

The metabolism and excretion of OCs vary according to their chemical structure. In general, P450s are considerably more efficient in metabolizing the lower chlorinated members. Indeed, in the case of PCBs, the replacement of hydrogen atoms in the biphenyl rings by several chlorine atoms decreases PCB metabolism by drug metabolizing enzymes and makes them persistent.

OCs that are metabolized by these enzymes are rapidly excreted in the urine and bile (Apostoli et al., 2003).

Due to their chemical structure and molecular dimension that are similar to those of 2,3,7,8-tetrachlorodibenzo-p-dixoin, some PCB isomers induce CYP enzymes (in particular, the CYP1A1 and CYP1B isoforms) through binding to an intracellular receptor protein complex, aryl hydrocarbon receptor (AhR) (Safe 1990, 1994). Microsomal enzyme induction may be measured in terms of the EROD activity in hepatic biopsies to detect enzymatic induction in PCB-exposed subjects (Whyte et al., 2000). The measurement of urinary D-glucaric acid (U-DGA) has also been proposed as a test for enzyme induction by halogenated aromatic hydrocarbons in humans. However, Apostoli et al. (2003) found no increase in U-DGA levels in human subjects with blood concentrations of aromatic hydrocarbons up to 394 µg/l, thereby raising doubts on the validity of these biomarker responses.

Mehmood et al. (1996) observed biotransformation of the pesticide HCB into PCP in the microsomal fractions of the whole cells of *Saccharomyces cerevisiae* expressing human CYP 3A4; PCP was further transformed into tetrachlorohydroquinone. Although the rate of metabolism was low, the results indicated that CYP 3A4 plays a role in the metabolism of chlorinated environmental contaminants.

In a dietary intake and faecal excretion study using five male subjects, (Juan et al., 2002) blood concentrations of PCBs were found to be largely controlled by the PCB concentrations in the body fat, and the short-term changes in uptake did not affect the PCB concentration in any particular tissue or organ. Moreover, studies showed that in humans, PCBs and other persistent organics present in various 'lipid depots' in the body (blood lipid, subcutaneous fat, visceral fat, etc.) partition between the depots fairly rapidly, i.e. over days/weeks (Wolfe et al., 1982; Brown and Lawton, 1984; Juan et al., 2002). Thus, the net absorption efficiency of OCs depends on factors such as an individual's body fat and balance (i.e. body fat index (BFI) and weight loss/gain), properties of the compound (i.e. susceptibility to metabolism) and long-term exposures.

Accumulation in Different Organs

POPs have been reported to occur in many tissues, organs and body fluids. In the context of the usage, human tissues are one of the most accurate and precise means of assessing exposure to environmental pollutants. POP concentrations are most commonly measured in blood, breast milk and adipose tissues. Apart from these, there were attempts to monitor human contamination by measuring the POP concentrations in hair and semen. Further, there were isolated attempts to measure POPs in placenta, endometrium, kidney, liver, urine, feces and meconium. The collection, analysis and interpretation of results obtained from different tissues and body fluids have their respective merits and demerits. However, in recent years, most of the data regarding quantification of POPs in human tissues and organs has been generally reported on a lipid weight basis, which enables comparison of the data.

Monitoring Studies

Allsopp and Johnston (2000) as well as the Global Report by UNEP Chemicals (2003) provided extensive data on the occurrence of POPs in the environment and wildlife as well as in humans from various regions of the world. In particular, the UNEP report documents regional assessments of all the 12 POPs in the form of 12 regional reports; the UNEP report contains extensive information on the environment, plants and animals, including POP concentrations in human tissues. Many studies were conducted on POP concentrations in human tissues and organs, and the data obtained were used for the assessment of potential toxic effects on various physiological and immunological processes. Meanwhile, the data also revealed several spatial and temporal variations in POPs.

Spatial Variations

Barakat (2004) provided a comprehensive review on the threats posed by PTS to the environment and human health in Egypt. Based on the data, he found that the reported OC concentrations in human milk samples from Egypt declined during the 1990s, consistent with the regulatory restrictions on the use of these compounds. The author found decreasing trends in dieldrin and DDTs when human milk samples from 1987 were compared with recent samples. Interestingly, high HCH concentrations were observed in recent milk samples, thus confirming their current usage.

In recent years, municipal open dumping sites in developing countries appear to be one of the major sources of OCs such as PCDDs and PCDFs based on the fact that intentional and unintentional low-temperature burning of plastics and municipal wastes can produce these toxic chemicals. Kunisue et al. (2004) analyzed human milk samples collected from women living near such sites in Asian developing countries, namely, India, Cambodia, Philippines and Vietnam, and compared the data with those from control sites. They found that the TEQ levels and concentrations of dioxins and related chemicals were much higher in milk samples obtained from women living near dumping sites in India than in those from the control sites, indicating the existence of significant pollution sources near the dumping sites in India. Global comparison of the data also showed that TEQs of these chemicals in the Indian milk samples are among the highest in the world (Fig. 7.2). In addition, these values were considerably higher in bovine milk samples collected near these sites, implicating bovine milk as a potential source of these chemicals to the nearby residents.

In a comparison between countries, Polder et al. (2003) found that the concentrations of HCB, HCHs and DDE were 2, 10 and 3 times higher, respectively in breast milk samples from Russia than in those from Norway, while the concentrations of sum PCBs and TEQs of *mono*- and *ortho*-substituted PCBs in samples from Russia were in the same range as those in Norway (Johansen et al., 1994). In human milk samples collected from Monchegrosk and Murmansk regions of Russia, Polder et al. (1998) found significant differences in the concentrations of HCH, CHLs and certain PCB congeners between the regions, but no geographic variation was observed in the concentrations of PCDDs, PCDFs and non-ortho PCBs.

Similarly, in the human milk samples collected from one urban and two rural areas in Java, Indonesia, Burke et al. (2003) could not find any statistical difference in the concentrations of all the pesticides among the three regions. This is consistent with the findings of Harris et al. (2001) who stated that the area of residence appeared to have little effect on residue concentrations in breast

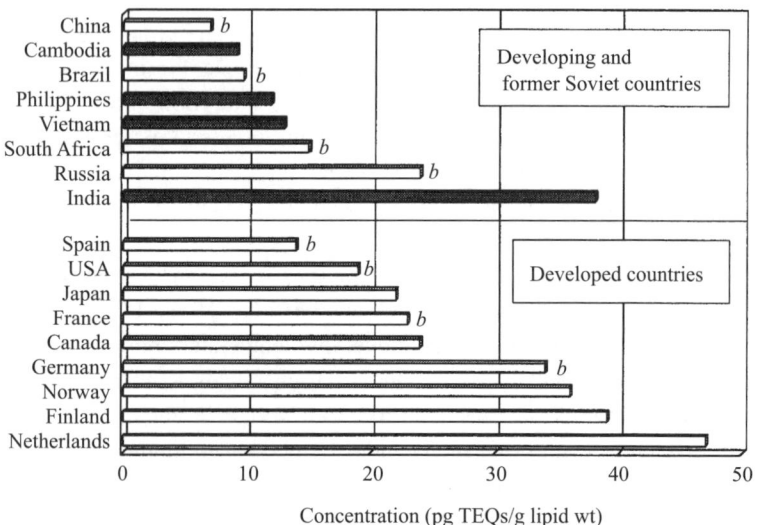

Fig. 7.2. Comparison of TEQ levels in human breast milk samples collected from women living near dumping sites in Asian developing countries with those collected from women in developed countries (Source: Kunisue et al., 2004).

milk even in areas where OCs were still commonly used. They also stated that unless women have been occupationally exposed (Kanja et al., 1986) or their homes have been treated with OCs (Quinsey et al., 1995), regional differences are not normally evident.

Paumgartten et al. (2000) found that the concentrations of PCDDs and PCDFs and TEQs in human milk samples from mothers living in Rio de Janeiro, Brazil, were lower than in those living in more industrialized countries such as Germany, France and Canada but were comparable to those living in countries such as China and Vietnam.

Recent findings by several authors indicated the suitability of human tissues, including breast milk, as indicators of spatial and regional differences in pollution by POPs in the ambient environment. Covaci et al. (2002) found that PCB profiles and concentrations in human adipose tissue samples from Belgium were similar to those observed in samples from other European countries such as Italy (Mariottini et al., 2000) and the UK (Davidson et al., 1994), indicating a similar PCB pollution profile in western European countries. Similarly, p,p'-DDE concentration in the Belgian population was similar to that in Swedish subjects (Meironyte et al., 2001). However, the concentrations of PCBs and DDTs were lower in human adipose samples from Belgium than in those from Chile and Mexico, respectively because of the continued usage of these chemicals in the latter countries. In a Mexican study on 287 human adipose tissue samples conducted over a 10-year period from 1988 to 1997, a higher DDT concentration was found in subjects from the suburban area of Veracruz, where DDT was used extensively to contain malaria-spreading vectors, than in subjects from the central of the town (Waliszewshi et al., 1998).

In human serum samples collected from Romania, Covaci et al. (2001) found regional differences in HCHs and DDTs; these were attributed to the differences in the age of the volunteers from whom the samples were collected. However, the authors found higher OC concentrations in human serum samples from Romanian volunteers than in samples from Belgian volunteers belonging to a similar age group.

As a part of the Flemish Environment and Health Study (FLEHS) conducted in Belgium, 47 pooled samples of human serum collected from 200 women aged between 50 and 65 years were analyzed for PCDDs, PCDFs and PCBs (Koppen et al., 2002). The concentrations of indicator PCBs and OC pesticides (HCHs, DDT, DDE, lindane and PCP) were found to be comparable to those found in samples from other European countries. Further, among the 47 pooled samples, PCB concentrations were higher in samples from urban areas than in those from rural areas, indicating the effect of urbanization and industrialization on PCB concentrations in humans.

While studying 411 human milk samples collected from five regions of Jordan, Nasir et al. (1998) found that samples from the North and Middle Valley regions were highly contaminated by DDTs and other pesticides when compared with those from Amman, Irbid and Zarqa regions.

Temporal Variations

A clear temporal trend was noticed in OC concentrations in breast milk samples obtained from women living in the Stockholm region during 1967–1997 (Noren and Meironyte, 2000). The study summarized the concentrations of PCBs, PCDDs, PCDFs, DDT, DDE, HCHs, dieldrin and some other persistent compounds. During the course of 20–30 years, the OC concentrations have decreased to varying extents. A 50% reduction in OC concentrations was achieved during a 4–17 year period. DDTs showed the most consistent decline; their half life was calculated to be 6 years (Fig. 7.3).

Another obvious temporal decrease in DDTs and total cyclodines was observed in human adipose tissues samples of Jordanian males and females during 1990–1996; this was consistent with the implementation of regulatory actions on the usage of these compounds. On the other hand, total HCH concentration increased marginally because of its continued usage in Jordan for the treatment of sheep and goat wool (Table 7.1) (Alawi et al., 1999). In Veracruz, Mexico, a statistically significant decrease in the DDT concentrations was observed during the survey years of 1988–1997 (Waliszewski et al., 1998).

Schade and Heinzow (1998) measured the concentrations of PCB, DDT and β-HCH in human milk samples collected from northern Germany over a 12-year period and followed the temporal trends. They found that compared with the values calculated 10 years earlier, the median DDT concentration decreased by 81%; PCBs, 61% and HCB and HCHs, 88% and 84%, respectively (Fig. 7.4).

Effect of Migration

Schmid et al. (1997) collected blood samples from male refugees with ages ranging from 16 to 60 years from Yugoslavia, USSR, Africa and Asia within

Table 7.1. Comparison of total concentrations of the main pesticide groups in adipose tissue samples taken from people of all ages in Jordan between 1990 and 1996 (Source: Alawi et al., 1999).

Group	1990 [1]	1996 (this paper)
Total DDT's	11.25	3.91
Total HCH's	1.22	1.55
Total Cyclodiens	4.00	0.804

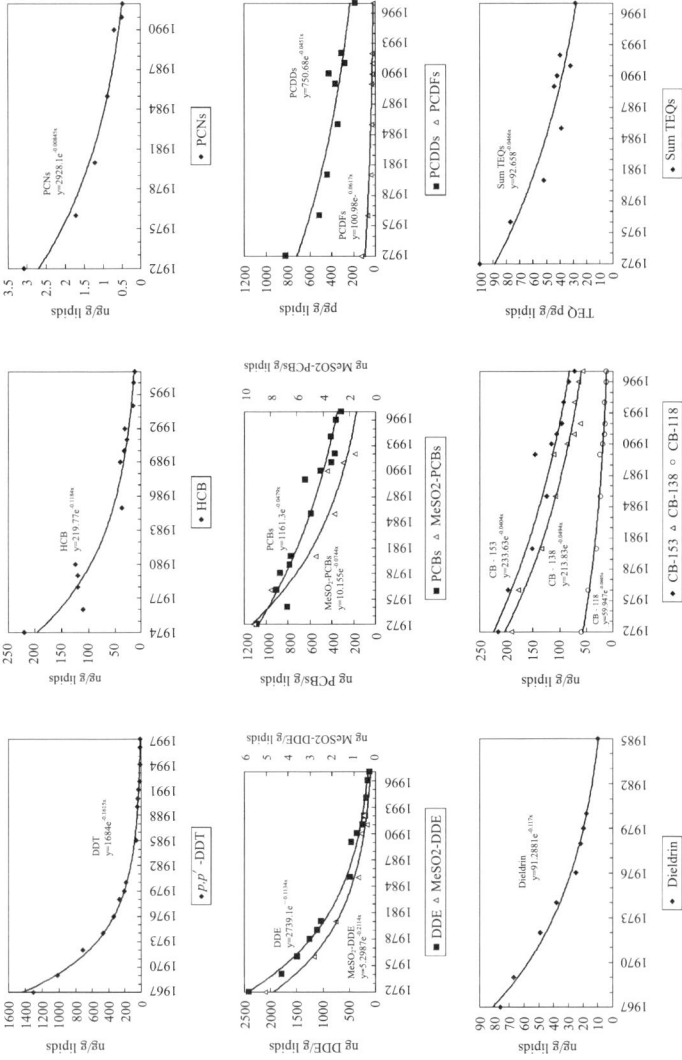

Fig. 7.3. Concentrations and temporal trends in p,p'-DDT, p,p'-DDE, MeSO₂-DDE, dieldrin, HCB, total PCBs, MeSO₂-PCBs, CB-153, CB-138, CB-118, PCNs, PCDDs, PCDFs and TEQs (sum of PCDDs, PCDFs and PCBs) in human milk expressed as exponential curves (Source: Noren and Meironyte, 2000).

20 days of their immigration to Germany from November 1994 to April 1995 and also from the residents of Germany (control group); the concentrations of DDE, PCBs, etc. in the plasma were quantified. They found a substantial regional difference among the various groups depending on their origin. The highest DDE concentrations were observed in Asians; PCBs, in Germans (Fig. 7.5). PCB concentrations in the plasma were below the detection limit in a majority of the refugees. β-HCH levels were the highest in refugees from Asia and the former USSR.

Koastsky et al. (1999) conducted an exploratory study on DDT and PCB concentrations in some Bangladeshi and Vietnamese immigrants settled in Canada. Apart from the effects of fish consumption, the authors observed that the concentrations of certain OCs, specifically PCBs, DDT, DDE and β-HCH, were the highest in the plasma of Bangladeshis than in the plasma of Vietnamese and the local people. They also found a correlation between the plasma DDT concentrations and the date of arrival of the Bangladeshi immigrants in Canada. With increase in the years of residency in Canada, there was a decrease in the DDT concentration in the plasma of Bangladeshis. As early as 1971, Radomski et al. (1971) observed that plasma β-HCH concentrations in Taiwanese students studying in the United States were 10 times higher than those in the US residents. Ip (1990) stated that the high concentrations of DDT and β-HCH in the breast milk samples of Hong Kong residents were due to the import of contaminated food from China.

Conclusions

Compared with wildlife, human organs, tissues and body fluids offer several advantages for use as bioindicators of POPs. Moreover, extrapolation of the results of toxicity studies conducted on other animals to understand the toxic responses in humans has its merits and demerits. Instead, the direct evaluation of human tissues may provide an accurate and easy method of assessment of human toxicity.

While collecting samples from human subjects, a well-planned and predetermined non-destructive sampling can be performed with utmost accuracy in volunteers after obtaining their informed consent. Ethical considerations may preclude sampling human tissues. However, appropriate institutional review board approvals are required. Moreover, essential details such as age, dietary habits and history of exposure to POPs can be easily and accurately obtained from the subjects. In the case that samples are obtained from hospitals during autopsies or biopsies, the above details can be gathered from the hospital records or from relatives. These data can be obtained with utmost accuracy only

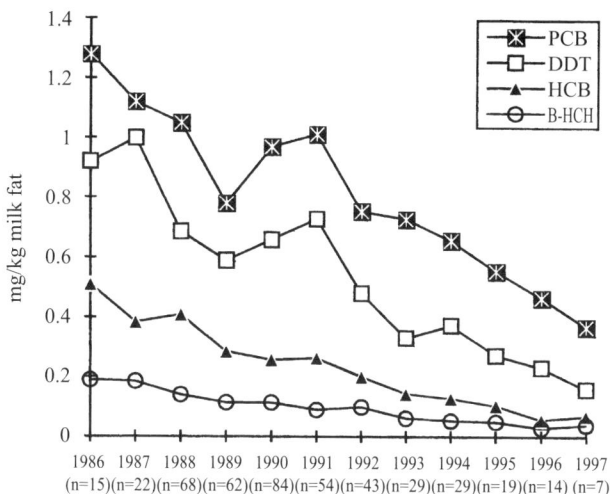

Fig. 7.4. Median concentrations of PCBs, DDT, HCB and β-HCH in homogenous subsamples collected between 1986 and 1997 (n = number of selected milk samples analyzed) (Source: Schade and Heinzow, 1998).

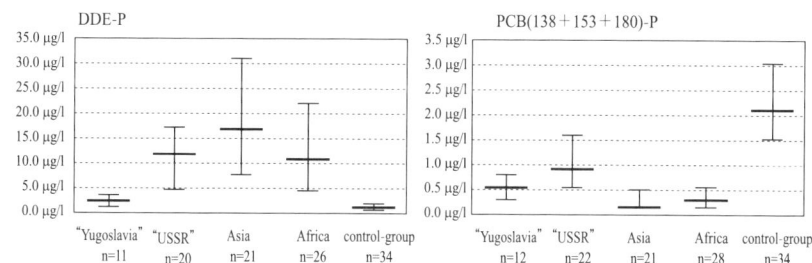

Fig. 7.5. Concentrations of DDE and PCBs in the plasma (lg/l) of different groups (median, 25th percentile and 75th percentile) (Source: Schmid et al., 1997).

from human subjects, but not wildlife. Therefore, despite the numerous legal, ethical, religious and social problems, human tissues represent one of the best biomonitors of POPs.

Generally, human milk, adipose tissue and blood serum have been used and found suitable as indicators of temporal and geographic variations in pollutants.

Apart from these samples, hair, semen, placenta and some other tissues have also been used for this purpose; however, these may not be suitable samples for POP monitoring because of their low lipid content and hence their low POP concentrations. Obtaining adipose tissue samples may be difficult because these can be obtained only from patients undergoing surgeries for various diseases, and the condition of the patient may indirectly affect the POP concentrations. Human breast milk is a very promising matrix for monitoring POP contamination between different communities, regions and even countries. It has been proved to be a very good indicator for evaluating temporal trends in POPs in different regions of the world. The major problem associated with these samples is that they can be obtained only from lactating females of the reproductive age. Therefore, when the extent of exposure or age-related POP accumulation is to be measured along with temporal trends and spatial variations in the entire population, blood samples from healthy males and females may prove suitable. Although POP concentrations may be lower in blood samples than in adipose tissue and human milk, the use of pooled samples from homogenous groups may provide a solution to this problem.

Apart from the legal formalities associated with the collection and transport of human tissues, organs and body fluids in different countries, several ethical, religious and social problems should be considered while collecting human samples, particularly in developing countries where a considerable proportion of the population is illiterate.

In the editorial section of the journal Marine Pollution Bulletin, Tanabe (2002) expressed his concern regarding a 'POPs contaminated future population', particularly in developed nations. He pointed out that because of late marriages and single child families in developed nations, the mother may transfer higher amounts of POPs, accumulated over the years prior to child bearing, to her single child; this may result in a population with high POP exposure. Future populations of developing nations will also be contaminated by POPs, albeit to a lesser extent. Recently, Dutch researchers found that babies who received high doses of dioxins *in utero* and through breast milk suffered reduced mental capacities (Tanabe, 2002). The toxins had impacts on the mental and psychomotor development of the children. These findings led the author to raise the question of whether this would lead to future research to ensure a suitable young work force 'without contamination' and also with proper 'physical and mental fitness'.

Hence, it may be advisable to collect human tissues, organs and body fluids, wherever and whenever necessary, both from developing and developed countries in order to track the extent of POP contamination and to save the future populations from its ill effects. Apart from the bioindicator organism(s) that may be recommended for continuous monitoring, human tissue samples may also be considered as one of the prime target areas of GEF/UNEP recommendation.

References

Ahlborg, U.G., L. Lipworth, L. Titus-Ernstoff, C.C. Hsieh, A. Hanberg, J. Baron, D. Trichopoulos and H.O. Adami, 1995. Organochlorine compounds in relation to breast cancer, endometrial cancer and endometriosis: an assessment for the biological and epidemiological evidence. Crit. Rev. Toxicol., 25: 463–531.

Alawi, M.A., S. Tamimi and M. Jaghabir, 1999. Storage of organochlorine pesticides in human adipose tissues of Jordanian males and females. Chemosphere, 38: 2865–2873.

Allsopp, M. and P. Johnston, 2000. Unseen Poisons in Asia. A review of persistent organic pollutant levels in South and Southeast Asia and Oceania. Exeter EX4 4PS, UK, p. 61.

Apostoli, P., A. Mangili, S. Carasi and M. Manno, 2003. Relationship between PCBs in blood and D-glucaric acid in urine. Toxicol. Lett., 144: 17–26.

Barakat, A.O., 2004. Assessment of persistent toxic substances in the environment of Egypt. Environ. Int., 30: 309–322.

Bazulic, D., B. Stampar-Plasaj, V. Bujanovic, N. Stojanovski, N. Nastev, I. Rudelic, N. Sisul and A. Zubeg, 1984. Organochlorine pesticide residues in the serum of mothers and their newborns from three Yugoslav towns. Bull. Environ. Contam. Toxicol., 32: 265–268.

Brown, J.F. and R.W. Lawton, 1984. Polychlorinated biphenyl (PCB) partitioning between adipose tissue and serum. Environ. Contam. Toxicol., 33: 277–280.

Burke, E.R., A.J. Holden and I.C. Shaw, 2003. A method to determine residue levels of persistent organochlorine pesticides in human milk from Indonesian women. Chemosphere, 50: 529–535.

Campoy, C., M.F. Olea-Serrano, M. Jimenez, R. Bayes, F. Canabate, M.J. Rosales, E. Blanca and N. Olea, 2001. Diet and organochlorine contaminants in women of reproductive age under 40 years old. Early Hum. Dev., 65 (Suppl.): S173–S182.

Chikuni, O., C.F.B. Nhachi, N.Z. Nyazema, A.A. Polder, I. Nafstad and J.U. Skaare, 1997. Assessment of environmental pollution by PCBs, DDT and its metabolites using human milk of mothers in Zimbabwe. Sci. Total Environ., 199: 183–190.

Covaci, A., C. Hura and P.S. Chepens, 2001. Selected persistent organochlorine pollutants in Romania. Sci. Total Environ., 280: 143–152.

Covaci, A., J. de Boer, J.J. Ryan, S. Voorspoels and P. Schepens, 2002. Distribution of organochlorinated contaminants in Belgian human adipose tissue. Environ. Res. Sect. A, 88: 210–218.

Cruz, S., C. Lino and M.I. Silveria, 2003. Evaluation of organochlorine pesticide residues in human serum from an urban and two rural populations in Portugal. Sci. Total Environ., 317: 23–35.

Davidson, R., S.C. Wilson and K.C. Jones, 1994. PCBs and other organo-chlorines in human tissues from the Welsh population: I—adipose. Environ. Pollut., 84: 69–77.

DeVoto, E., L. Kohlmeier and W. Heeschen, 1998. Some dietary predictors of plasma organochlorine concentrations in an elderly German population. Arch. Environ. Health, 53: 147–155.

Foster, W.G., C.L. Hughes, S. Chan and L. Platt, 2002. Human developmental exposure to endocrine active compounds. Environ. Toxicol. Pharmacol., 12: 75–81.

Gallelli, G., S. Mangini and C. Gerbino, 1995. Organochlorine residues in human adipose and hepatic tissues from autopsy sources in northern Italy. J. Toxicol. Environ. Health, 46: 293–300.

Hanaoka, T., Y. Takahashi, M. Kobayashi, S. Sasaki, M. Usuda, S. Okubo, M. Hayashi and S. Tsugane, 2002. Residues of beta-hexachlorocyclohexane, dichlorodiphenyl-trichloroethane and hexachlorobenzene in serum and relations with consumption of dietary components in rural residents in Japan. Sci. Total Environ., 286: 119–127.

Harris, C.A., S. O'Hagen and G.H.J. Merson, 2001. Factors affecting transfer of organochlorine pesticide residues to breast milk. Chemosphere, 43: 243–256.

Hura, C., M. Leanca, L. Rusu and B.A. Hura, 1999. Risk assessment of pollution with pesticides in food in the Eastern Romania area (1996–1997). Toxicol. Lett., 107: 103–107.

Ip, H.M.H., 1990. Chlorinated pesticides in foodstuffs in Hong Kong. Arch. Environ. Contam. Toxicol., 19: 291–296.

Johansen, H.R., G. Becher, A. Polder and J.U. Skaare, 1994. Congener-specific determination of polychlorinated biphenyls and organochlorine pesticides in human milk from Norwegian mothers living in Oslo. J. Toxicol. Environ. Health, 25: 1–19.

Juan, C.Y., G.O. Thomas, A.J. Sweetman and K.C. Jones, 2002. An input-output study for PCBs in humans. Environ. Int., 28: 203–214.

Kang, Y., M. Matsuda, M. Kawano, T. Wakimoto and B. Min, 1997. Organochlorine pesticides, polychlorinated biphenyls, polychlorinated dibenzo-*p*-dioxins and dibenzofurans in human adipose tissue from western Kyungnam, Korea. Chemosphere, 35: 2107–2117.

Kanja, L.W., J.U. Skaare, I. Nafstd, C.K. Maitai and P. Lkken, 1986. Organochlorine pesticides in human milk from different areas of Kenya 1983–1985. J. Toxicol. Environ. Health, 19: 449–464.

Kannan, K., S. Tanabe, A. Ramesh, A.N. Subramanian and R. Tatsukawa, 1992. Persistent organochlorine residues in foodstuffs from India and their implications on human dietary exposure. J. Agric. Food Chem., 40: 518–524.

Kannan, K., S. Tanabe, R.J. Williams and R. Tatsukawa, 1994. Organochlorine residues in foodstuffs from Australia, Papua New Guinea and Solomon islands: contamination levels and human dietary exposure. Sci. Total Environ., 153: 29–49.

Kashyap, R., L.R. Iyer, M.M. Singh and S.K. Kashyap, 1994. Assessment of location-specific human exposure to dichlorodiphenyltrichloroethane and benzenehexachloride in Gujarat State, India. Environ. Health, 65: 381–384.

Koastsky, T., R. Przybysz, B. Shatenstein, J.P. Weber and B. Armstrong, 1999. Contaminant exposure in Montrealers of Asian origin fishing the St. Lawrence River: Exploratory assessment. Environ. Res. Sect. A, 80: S159–S165.

Koppen, G., A. Covaci, R. van Cleuvenbergen, P. Schepens, G. Winneke, V. Nelen, R. van Larebeke, R. Vlietinck and G. Schoeters, 2002. Persistent organochlorine pollutants in human serum of 50–65 years old women in the Flanders Environmental and Health Study (FLEHS). Part 1: concentrations and regional differences. Chemosphere, 48: 811–825.

Kunisue, T., M. Someya, I. Monirith, M. Watanabe, T.S. Tana and S. Tanabe, 2004. Occurrence of PCBs, organochlorine insecticides, tris(4-chloro-phenyl)methane and tris(4-chlorophenyl)methanol in human breast milk collected from Cambodia. Arch. Environ. Contam. Toxicol., 46: 405–412.

Letz, G., R. Wabeke and R. Weinstein, 1990. Polychlorinated biphenyl (PCB) toxicity. J. Toxicol.: Clin. Toxicol., 28: 505–526.

Li, Y.F., D.J. Cai and A. Singh, 1998. Technical hexachlorocyclohexane use trends in China and their impact on the environment. Arch Environ. Contam. Toxicol., 35: 688–697.

Loffler, G. and B. van Bavel, 2000. Potential pathways and exposure to explain the human body burden of organochlorine compounds: a multivariate statistical analysis of human monitoring in Wurzburg, Germany. Chemosphere, 40: 1075–1082.

Manirakiza, P., O. Akimbamijo, A. Covaci, S.A. Adediran, I. Cisse, S.T. Fall and P. Schepens, 2002. Persistent chlorinated pesticides in fish and cattle fat and their implications for human serum concentrations from the Sene-Gambrain region. J. Environ. Monitor., 4: 609–617.

Mariottini, M., S. Aurigi and S. Focardi, 2000. Congener profile and toxicity assessment of polychlorinated biphenyls in human adipose tissues of Italians and Chileans. Microchem. J., 67: 63–71.

Marquardt, H. and S. Schafer, 1994. (Eds.). Lehrbuch der Toxikologie. BI-Wiss-Verl., Mannheim, Leipzig, Wien, Zurich, p. 440.

Mehmood, Z., M.P. Williamson, D.E. Kelly and S.L. Kelly, 1996. Metabolism of organochlorine pesticides: The role of human cytochrome P450 3A4. Chemosphere, 33: 759–769.

Meironyte, D., A. Bergman and K. Noren, 2001. Polybrominated diphenyl ethers in Swedish human liver and adipose tissue. Arch. Environ. Contam. Toxicol., 40: 564–570.

Minh, T.B., M. Watanabe, S. Tanabe, T. Yamada, J. Hata and S. Watanabe, 2001. Specific accumulation and elimination kinetics of tris(4-chlorophenyl)methane, tris(4-chlorophenyl)methanol and other persistent organochlorines in humans from Japan. Environ. Health Perspect., 109: 927–935.

Nakata, H., S. Tanabe, R. Tatsukawa, M. Amano, N. Miyazaki and E.A. Petrov, 1995. Persistent organochlorine residues and their accumulation kinetics in Baikal seal (Phoca sibirica) from Lake Baikal. Environ. Sci. Technol., 29: 2877–2885.

Nakata, H., S. Tanabe, R. Tatsukawa, Y. Koyama, N. Miyazaki, S. Belikov and A. Boltunov, 1998. Persistent organochlorine contaminations in ringed seals (Phoca hispida) from the Kara Sea, Russian Arctic. Environ. Toxicol. Chem., 17: 1745–1755.

Nakata, H., M. Kawazoe, K. Arizono, S. Abe, T. Kitano, H. Shimada, W. Li and X. Ding, 2002. Organochlorine pesticides and polychlorinated biphenyl residues in foodstuffs and human tissues from China: Status of contamination, historical trend and human dietary exposure. Arch. Environ. Contam. Toxicol., 43: 473–480.

Nasir, K., Y.Y. Bilto and Y. Al-Shuraiki, 1998. Residues of chlorinated hydrocarbon insecticides in human milk of Jordanian women. Environ. Pollut., 99: 141–148.

Nicholson, W.J. and P.J. Landrigan, 1994. Human health effects of polychlorinated biphenyls. In: A. Schecter (Ed.), Dioxins and Health. Plenum Press, New York, pp. 487–525.

Noren, K., 1983. Some aspects of determination of organochlorine contaminants in human milk. Arch. Environ. Contam. Toxicol., 12: 277–283.

Noren, K. and D. Meironyte, 2000. Certain organochlorine and organobromine contaminants in Swedish human milk in perspective of past 20–30 years. Chemosphere, 40: 1111–1123.

Paumgartten, F.J.R., C.M. Cruz, I. Chahoud, R. Palavinskas and W. Mathar, 2000. PCDDs, PCDFs, PCBs and other organochlorine compounds from human milk from Rio de Janeiro, Brazil. Environ. Health Sect. A, 83: 293–297.

Pauwels, A., A. Covaci, J. Weyler, L. Delbeke, M. Dhont, P. De Sutter, T. D'Hooghe and P. Schepens, 2000. Comparison of persistent organic pollutant residues in serum and adipose tissue in a female population in Belgium, 1996–1998. Arch. Environ. Contam. Toxicol., 39: 265–270.

Polder, A., G. Becher, T.N. Svinova and J.U. Skaare, 1998. Dioxins and some chlorinated pesticides in human milk from the Kola Peninsula, Russia. Chemosphere, 37: 1795–1806.

Polder, A., J.O. Odland, A. Tkachev, S. Foreid, T.N. Savinova and J.U. Skaare, 2003. Geographic variation of chlorinated pesticides, toxaphenes and PCBs in human milk from sub-arctic locations in Russia. Sci. Total Environ., 306: 179–195.

Quinsey, P.M., D.C. Donohue and J.T. Ahokas, 1995. Persistence of organo-chlorines in breast milk of women in Victoria, Australia. Food Chem. Toxicol., 33: 49–56.

Radomski, J.L., W.B. Deichman and W.B. Rey, 1971. Human pesticide blood levels as a measure of body burden and pesticide exposure. Toxicol. Appl. Pharmacol., 20: 175–185.

Safe, S., 1990. Polychlorinated biphenyls (PCBs), dibenzo-p-dioxins (PCDDs), dibenzofurans (PCDFs) and related compounds: environmental and mechanistic considerations which support the development of toxic equivalency factors (TEFs). Crit. Rev. Toxicol., 21: 51–88.

Safe, S., 1994. Polychlorinated biphenyls (PCBs): environmental impact, biochemical and toxic responses and implications for risk assessment. Crit. Rev. Toxicol., 24: 87–149.

Saxena, M.C., T.D. Seth and P.L. Mahajan, 1980. Organochlorine pesticides in human placenta and accompanying fluid. Int. J. Environ. Anal. Chem., 7: 248–251.

Schade, G. and B. Heinzow, 1998. Organochlorine pesticides and polychlor-inated biphenyls in human milk of mothers living in northern Germany: current extent of contamination, time trend from 1986 to 1997 and factors that influence the levels of contamination. Sci. Total Environ., 215: 31–39.

Schaefer, W.R., B.S. Hermann, I. Meinhold-Heerlein, W.R. Deppert and H.P. Zahradnik, 2000. Exposure of human endometrium to environmental estrogens, antiandrogens and organochlorine compounds. Fertil. Steril., 74: 558–563.

Schmid, K., P. Lederer, T. Goen, K.H. Schaller, H. Strebl, A. Weber, J. Angerer and G. Lehnert, 1997. Internal exposure to hazardous substances of persons from various countries: investigations on exposure to different organochlorine compounds. Int. Arch. Occup. Environ. Health, 69: 399–406.

Siddiqui, M.K.J., M.C. Saxena and C.R. Krishna Murti, 1981. Storage of DDT and BHC in adipose tissue of Indian males. Int. J. Environ. Anal. Chem., 10: 197–204.

Smeds, A. and P. Saukko, 2001. Identification and quantification of polychlorinated biphenyls and some endocrine disrupting pesticides in human adipose tissue from Finland. Chemosphere, 44: 1463–1471.

Smith, A.G. and S.D. Gangolli, 2002. Organochlorine chemicals in seafood: occurrence and health concerns. Food Chem. Toxicol., 40: 767–779.

Stuetz, W., T. Prapamontol, J.G. Erhardt and H.G. Classen, 2001. Organochlorine pesticide residues in human milk of a Hmong hill tribe living in northern Thailand. Sci. Total Environ., 273: 53–60.

Subramanian, A.N., S. Tanabe, Y. Fujise and R. Tatsukawa, 1986. Organochlorine residues in Dall's and True's porpoises collected from northwestern Pacific and adjacent waters. Mem. Natl. Inst. Polar Res., 44: 167–173.

Tanabe, S., 2000. Asian developing regions: persistent organic pollutants in the seas. In: C.R.C. Sheppard, (Ed.), Seas at the Millennium: An Environmental Evaluation, Pergamon, Amsterdam, The Netherlands, pp. 447–462.

Tanabe, S., 2002. Higher contamination in the future population in developed nations. Mar. Pollut. Bull., 44: 1315–1316.

Tanabe, S., F. Gondaira, A.N. Subramanian, A. Ramesh, D. Mohan, P.L. Kumaran, V.K. Venugopalan and R. Tatsukawa, 1990. Specific pattern of persistent organochlorine residues in human breast milk from South India. J. Agric. Food Chem., 38: 899–903.

UNEP Chemicals, 2003. Regionally Based Assessment of Persistent Toxic Substances. Global Report 2003, p. 207, http: //www.chem.unep.ch/pts.

Waliszewski, S.M., A.A. Aguirre, R.M. Infanzon, J. Rivera and R. Infanzon, 1998. Time trend of organochlorine pesticide residues in human adipose tissue in Veracruz, Mexico: 1989–1997 survey. Sci. Total Environ., 221: 201–204.

Wolfe, M.S., J.T. Thronton, A. Fischbein, R. Lilis and I.J. Selikoff, 1982. Disposition of polychlorinated biphenyl congeners in occupationally exposed persons. Toxicol. Appl. Pharmacol., 62: 294–306.

Whyte, J.J., R.E. Jung, C.J. Schmidt and D.E. Tillitt, 2000. Ethoxyresorufin-*O*-

deethylase (EROD) activity as a biomarker of chemical exposure. Crit. Rev. Toxicol., 30: 347–370.

Yoshida, K., S. Ikeda and J. Nakanishi, 2000. Assessment of human health risks of dioxins in Japan. Chemosphere, 40: 177–185.

Mussels: Preferred as Bioindicators of POPs over other Animals

Mussels are always preferred as bioindicators for many pollutants because of their ability to accumulate a myriad of contaminants. They can withstand varying environmental conditions and bioaccumulate contaminants to higher levels through filter-feeding. They also possess all the characteristics required of a good bioindicator of POPs.

Chapter 8: Conclusions and Recommendations

An Ongoing Challenge

POPs were first used more than five decades ago; this was soon followed by recognition of their persistence in the environment as well as biotopes. More than 50 years have passed since the publication of the first reports on the environmental contamination and accumulation of POPs in non-target organisms. Several attempts have been made to curtail and/or reduce the contamination of the biota and environment by these POPs. Despite the regulatory steps taken by many nations, POPs continue to pose the same challenge as 50 years earlier. The only positive result of such efforts by the international community is the decrease in the concentrations of the earliest used POPs such as DDTs in animal bodies. With the exception of such sporadic events, the scientific community is still in the initial stage with regard to abating the ill-effects of these chemicals on the environment and biota, including humans.

Several legislations have been passed by many nations, and guideline values have been proposed by many national and international agencies such as the WHO, FAO, UNEP, US EPA, Environment Canada and the Japanese Ministry of Environment. Several attempts have been made to conserve and enhance the ecological health of the world's freshwater and marine ecosystems—rivers, streams, lakes, wetlands, estuaries, coastal waters, oceans and ground water—to support healthy aquatic communities in future through the availability of safe water. In the past few decades, guidelines for the use and control of persistent chemicals in terrestrial and aquatic environments have been streamlined and enforced to varying degrees in different nations.

Despite these efforts, POPs have become ubiquitous in both biotic and abiotic spheres. POPs continue to pose the same challenges, and several questions regarding methods of obtaining a clean environment remain unanswered. Answering critical questions regarding the physical, chemical and biological state of the world's waters and the toxic effects of POPs on biota remains a challenge. One of the most meaningful ways of answering these questions is to directly observe the plants and animals that live in the water bodies because these biological indicators, that is, aquatic organisms and their communities, integrate the effects of various stressors and reflect the current conditions as well as changes occurring over time and cumulative effects. Biological indicators can help identify problems that may be otherwise missed or underestimated. Thus, assessments through bioindicators can be an important component of watershed management programs. They provide integrated evaluations of water quality. Resident biota can function as continual monitors of environmental quality, increasing the likelihood of detecting the effects of episodic events or other impacts over time, which is unlikely with periodic chemical sampling. The main advantage of bioindicators is that they can accumulate POPs to many folds higher than those in the ambient environment; this may be missed or underestimated in direct measurements. Such biological estimates can also be used to determine the extent to which current regulations are protecting the designated environment and the life forms living in the region.

Animals belonging to all orders, comprising many genera, have been analyzed for their POP content for understanding the concentrations, toxic effects and metabolism of these chemicals. Several attempts were also made to utilize many of these animals as bioindicators of exposure to POPs. Apart from the six groups of animals discussed in the previous six chapters of this review, limited literature is available regarding other genera. In spite of the moderate amount of information available regarding certain animals, they have not been useful as bioindicators of POPs due to various reasons such as biological, social, ethical, legal and logistical problems.

Many plants were assessed as accumulators of POPs. Many plants and animals, ranging from planktonic organisms to humans, have been analyzed for their POP content. It may be interesting to cite some of these organisms that have not been considered in previous chapters. For example, the effects of pesticide toxicity and their distribution in planktonic organisms have been reviewed by Hanazato (2001). Takeuchi et al. (2001) have proposed 'Caprella Watch' that includes the use of *Caprella* sp. to monitor temporal and spatial changes in baseline concentrations of butyltins.

A vast amount of literature is available on the accumulation characteristics of POPs in terrestrial and aquatic plants. Although vegetation has a great potential to accumulate organic pollutants, little is known regarding the quantitative importance of vegetation as a sink for these chemicals (Simonich and Hites, 1995). Davidson et al. (2003) analyzed conifer needles from mountain plants at different altitudes for HCHs, DDTs and PCBs and found that the more volatile OCs occurred in higher concentrations at higher altitudes, whereas the occurrence of less volatile OCs was either unrelated or inversely related with altitude. This suggests that some plants may be used for determining the local spatial distribution of POPs. Kylin and Sjdin (2003) recently found biological and site-related variations in HCHs and DDTs in Scots pine needles. The stomatal pathway involved in the exchange of PCBs between air and plants was evaluated by Barber et al. (2002), and the uptake and translocation of CHL in food crops was studied by Mattina et al. (2000). Uptake of airborne PCDDs and PCDFs by a native pasture sward was evaluated by Thomas et al. (2002). Kelley and Gobas (2001) estimated the uptake of POPs by lichens collected from the Arctic. DDT, HCB and PCBs present in mango leaf samples from West Africa were quantified by Bacci et al. (1988). Haynes et al. (2000) examined seagrasses from Australia for their concentrations of dieldrin, DDTs and HCHs as well as some other pesticides and herbicides. Several Egyptian spice and medicinal plants (Abou-Arab and Abou Donia, 2001) and sunflower seeds (Prados-Rosales et al., 2003) were also tested for their POP concentrations.

In addition, terrestrial and aquatic organisms such as earthworms (Gevao et al., 2001), adult male rats (You et al., 2001) and marine sponges (Perez et al., 2003) as well as turtle eggs (de Solla et al., 2001) and semen of farm animals (Kamarianos et al., (2003) were also analyzed for determining their concentrations of various POPs; however, many of these reports are sporadic.

Conclusions

Based on the information presented in earlier chapters, we found that six groups of organisms—mussels, squids, fish, birds, marine mammals and humans—can provide necessary information on the POP pollution status in their respective environments.

To summarize, the conformity of the above six groups of animals to the basic indicator prerequisites, which were explained in Chapter 1 and are essential for widespread monitoring of POPs, is matched by their marked ability to act as bioindicator organisms. Research on POP monitoring using these organisms has been very common.

Mussels
Bivalve mollusks appear to be the best bioindicators for regular monitoring of POPs in developing countries due to several reasons explained earlier. They have been extensively used with success to assess the levels and trends of several POPs. Many bivalve species, particularly mussels and oysters, appear to be highly suitable for use as bioindicators. In particular, species belonging to the genera *Perna*, *Mytilus* and *Crassotrea*, in that order of preference, can be gathered and evaluated.

However, it is important to note that no organism is a perfect indicator of pollution, and the use of bivalves to monitor POPs is also subject to various

Table 8.1. Comparison of characteristics of different groups of animals for their suitability as bioindicators of pollution by POPs.

No.	Characteristics*	Mussel	Squid	Fish	Bird	Marine mammal	Human
1.	Measurable response	O	O	O	×	×	×
2.	Territory overlapping	O	△	△	×	×	△
3.	Easy sampling	O	O	O	×	×	△
4.	Easy handling	O	O	O	△	×	△
5.	Enough availability	O	O	O	×	×	×
6.	Optimum life span	△	×	△	O	O	O
7.	Simple feeding habit	O	O	△	×	×	×
8.	Sedentary life style	O	△	△	△	×	△
9.	Withstand extreme conditions	O	×	×	O	O	O
10.	Taxonomically well known	O	O	O	O	O	O
11.	Well known biology and life history	O	△	△	△	△	O
12.	Possibilities of easy survey	O	△	△	×	×	O
13.	Broad geographical range	O	×	△	△	△	O
14.	Extrapolatable results	O	△	△	△	△	△
15.	Economic importance	O	O	O	×	×	×
16.	Cost-effective	O	O	O	×	×	×
17.	Internationally transportable	O	O	O	×	×	×

*: For explanation on characteristics refer Chapter 1.
O: Suitable
△: Depend on variables (species)
×: Not suitable

confounding factors. Variations that may occur due to season, size, sampling location, etc. can be very easily overcome with bivalves than with other organisms. All the 17 characteristics, which have been explained in Chapter 1, may be particularly important for an organism to be considered a suitable bioindicator for POP monitoring. Based on the 17 characteristics, all the six groups of organisms that have been discussed in the previous chapters are ranked in Table 8.1.

Squids

Existing reports testify that squids are possible bioindicator organisms for POP monitoring in shallow coastal environments. Reviews of available literature have revealed that the OCs in squid specimens reflect their levels in the ambient environment, and seasonal and growth-related factors do not affect the OC concentrations in squids. The numerous squid species available at any given location renders the task of selecting the indicator species easier. As stated earlier, the problems associated with using squids as indicators of POPs are their very short lifespan and their mobility. However, if the need arises, squids may also be used as bioindicators of pollution by POPs.

Fish

Pollutant kinetics in fish has been the subject of widespread research, and there have been numerous efforts to use fish as monitors of POPs. The most important reason for this emphasis is undoubtedly their economic importance as well as easy availability. However, problems do exist. Mobility is a major problem in the use of finfish as bioindicators of POPs. Many authors have ignored this problem, assuming that the POP concentrations present in their tissues are characteristic of the site at which they were caught. Many coastal species migrate into estuaries or near-shore waters during different stages of their lives. This problem can be partly overcome by the use of demersal fish species that have a limited range of migration.

Birds

Birds can very well be used as biomonitors of pollution by POPs worldwide. They can not only reflect the POP levels in their habitat but also be used as monitors of global contamination by POPs on account of their migratory nature. However, if and when birds are used as bioindicators of POPs, their trophic level in the food chain, their metabolic capacities for POPs, geographical distribution, gender differences, migration and ethical and legal problems related to the collection and transport of the specimens should be taken into consideration.

Marine Mammals

Although marine mammals are very good bioaccumulators of all POPs and are useful bioindicators for measuring long-term temporal trends and spatial variations, several factors may interfere in their selection as indicators of POPs. They may not be good indicators of pollution by POPs because of their low susceptibility to short-term changes in pollution. Moreover, it may be difficult to obtain suitable and adequate numbers of tissue samples for regular monitoring of POPs.

Many species of cetaceans are included in the CITES, and their transport between countries are restricted by various regulations. Moreover, the use of these animals has several limitations such as enormous body size, difficulties in dissection and obtaining sub-samples of various tissues and organs, wide migrations, large male-female differences in residue levels, varying metabolic capacities for POPs among species and different life-stages of the same species. Regardless of all these limitations, there is no doubt that marine mammals, both cetaceans and pinnipeds, may be good indicators for studying integrated temporal trends in POPs in the marine environments, particularly the oceans.

Humans

The use of human organs, tissues and body fluids as bioindicators of POPs has several advantages over the use of other animals. Well-planned and predetermined non-destructive sampling can be performed with utmost accuracy after obtaining informed consent from volunteers. However, the biggest predicaments in sampling human tissues and body fluids are the ethical, social, religious and legal problems. Adipose tissues, which are good accumulators of POPs, cannot be easily obtained; breast milk samples can be obtained only from lactating females in the reproductive age group; semen can be obtained only from males; and hair and blood samples may pose problems during analysis due to their very low POP content. However, as has been foreseen by several authors, human problems anticipated with POP usage dictate the necessity of monitoring of contamination by POPs to protect the future generations from the ill effects of POPs. Apart from the bioindicator organism(s) that may be recommended for continuous monitoring, analysis of POPs in human tissues can also be considered as a prime focus area of GEF research on POPs.

Recommendations

Table 8.1 clearly shows that with the use of a well-planned sampling and analysis schedule, bivalve mollusks may be the best sentinel organisms in aquatic environments for measuring the extent of pollution in the adjacent terrestrial regions of developing countries and can contribute to the global database on the distribution of POPs in the environment.

With certain precautions, bivalve mollusks can be considered as near-perfect sentinel organisms. A comparison of the biology of various species and pollutant kinetics of POPs in these organisms shows that bivalves have a clear potential as bioindicators of these chemicals. As shown in Table 8.1, most bivalve species conform well to the basic indicator prerequisites. Bivalves contain high POP concentrations (at least of the most common POPs such as DDTs, PCBs and CHLs) when compared with those in most other aquatic animals at the same trophic level, and there is little evidence of metabolism or regulation of most chemicals in bivalve species.

Mollusks have certainly been the most commonly used bioindicators of many pollutants, including POPs. However, before using bivalves for monitoring POPs in developing countries, refinement of the indicator techniques using bivalve species assumes great importance, particularly for monitoring on a global scale.

Collecting the same species or closely related species of bivalves from the entire study area is advised. We recommend mussels of the genus *Perna*, particularly the green mussel *Perna viridis*, for monitoring POPs in developing Asian countries, most of which are situated in the tropical belt where the species belonging to the genus *Perna* are most commonly found. Other mussel species of the same genus (e.g. *Perna indica, Perna picta* and *Perna vulsella*) may be used in places where green mussels are scarce.

It will always be better to collect bivalves of two or more genera common to the sampling location. For this purpose, species belonging to the genera *Crassostrea* (e.g. *Crassostrea virginica, Crassostrea gigas* and *Crassostrea madrasensis*) and *Mytilus* (e.g. *Mytilus edulis, Mytilus mytilus* and *Mytilus galloprovincialis*) can be collected. These species are preferred because they can withstand extremely severe environmental conditions. For comparison with these bivalves, other species of mussels, oysters and clams, which are commonly found in the coastal regions of tropical developing countries, may also be collected.

Parameters such as salinity and water temperature should also be taken into consideration, and it is preferable to collect specimens from coastal regions rather than from estuaries and backwaters to avoid excessive pollution and salinity and temperature variations. Areas polluted by other types should also be avoided because bivalves close their valves during extreme conditions and thus may not accumulate pollutants. As far as possible, same-sized specimens should be collected from all sampling sites. Spent specimens should certainly be avoided. Therefore, it is always preferable to follow the seasonal changes in specimens at each station and to fix the dates of collection after a general discussion among the group.

Field collection of repetitive samples is an extremely complex procedure, and various logistical parameters should be pre-planned; this necessitates the presence of technically skilled and scientifically competent field scientists at all sampling points. To achieve this, training workshops may be conducted by trained personnel in various regions.

If the need arises, apart from bivalves, other groups of animals discussed in this review, namely, fish, squids, birds and marine mammals, may also be used as bioindicators for assessing the levels of contamination by POPs in their ambient environments. In addition, analysis of human samples may also be treated as one of the prime focus areas of research on these persistent chemicals.

The trained field scientist must personally collect the samples or personally supervise the collection. Competent research centers in each country may be selected and given the responsibility of sampling, preservation and transport of specimens. This is extremely crucial because knowledge of local conditions, language and support are indispensable for such sampling in remote coastal areas. The research centers should be provided with all the necessary facilities such as GPS and freezer facilities as well as other logistical support.

Capacity building should be an integral part of the sampling and analytical program. If the analyses are to be conducted at regional centres, inter-laboratory comparison should be conducted prior to the start of the exercise. Depending upon the local conditions, the field and laboratory scientists must be efficient in crisis management.

Aquatic animals, particularly the coastal species, may form very good bioindicator species because of the several reasons provided in the previous chapters. In general, for regular monitoring of POPs, aquatic species may be the preferred organisms due to the following reasons.

Regardless of their nature and location of use, all chemicals finally reach the marine environment—the final reservoir—via various pathways.

The coastal ecosystems adjacent to the terrestrial environment easily reflect the changes in the levels of pollutants at their source.

Coastal ecosystems are also home to many species, including phytoplankton, zooplankton, aquatic plants, insects, fish, birds and mammals. Thus, they provide a wide choice for selection of bioindicators. These ecosystems are organized at many levels, encompassing communities, populations, species and genetic levels. Therefore, they include wide biodiversity and habitats when compared with terrestrial ecosystems.

Compared to terrestrial species, the aquatic species are in more intimate contact with their ambient environment. Therefore, the latter reflect environmental changes more accurately than their terrestrial counterparts.

The aquatic environment is more dynamic and homogenous in nature when compared with the terrestrial environment; hence, aquatic animals reflect even short-term changes in environmental pollutant levels in their body concentrations.

It may be possible to find large numbers of sedentary and semi-sessile organisms for pollutant analysis of the aquatic environment; these aquatic organisms, rather than terrestrial organisms, may truly reflect the changes in pollutant usage pattern(s) in the nearby terrestrial environment.

References

Abou-Arab, A.A.K. and M.A. Abou Donia, 2001. Pesticide residues in some Egyptian spices and medicinal plants as affected by processing. Food Chem., 72: 439–445.
Bacci, E., D. Calamari, C. Gaggi, C. Biney, S. Focardi and M. Morosini, 1988.

Organochlorine pesticide and PCB residues in plant foliage. Chemosphere,
17: 693–702.

Barber, J.L., G.O. Thomas, G. Kerstiens and K.C. Jones, 2002. Air-side and
plant-side resistances influence the uptake of airborne PCBs by evergreen
plants. Environ. Sci. Technol., 36: 3224–3229.

Davidson, D.A., A.C. Wilkinson, J.M. Blais, L.E. Kimpe, K.C. McDonald and
D.W. Schindler, 2003. Orographic cold-trapping of persistent pollutants
by vegetation in mountains of western Canada. Environ. Sci. Technol.,
37: 209–216.

de Solla, S.R., C.A. Bishop, H. Lickers and K. Jock, 2001. Organochlorine
pesticides, PCBs, diphenyldioxin and furan concentrations in common
snapping turtle eggs (*Chelydra serpentine serpentine*) in Akwesasne,
Mohawk territory, Canada. Arch. Environ. Contam. Toxicol., 40: 410–
417.

Gevao, B., C. Mordaunt, K.T. Semple, T.G. Piearce and K.C. Jones, 2001. Bioavailability of nonextractable (bound) pesticide residues to earthworms.
Environ. Sci. Technol., 35: 501–507.

Hanazato, T., 2001. Pesticide effects on freshwater zooplankton: an ecological
perspective. Environ. Pollut., 112: 1–10.

Haynes, D., J. Muller and S. Cartens, 2000. Pesticide and herbicide residues in
sediments and seagrasses from the Great Barrier Reef world heritage area
and Queensland coast. Mar. Pollut. Bull., 41: 279–287.

Kamarianos, A., X. Karamanlis, E. Theodosiadou, P. Goulas and A. Smokovitis,
2003. The presence of environmental pollutants in the semen of farm
animals (bull, ram, goat and boar). Reprod. Toxicol., 17: 439–445.

Kelley, B.C. and F.A.P.C. Gobas, 2001. Bioaccumulation of persistent organic
pollutants in lichen-caribou-wolf food chains of Canada's central and
western Arctic. Environ. Sci. Technol., 35: 325–334.

Kylin, H. and A. Sjodin, 2003. Accumulation of airborne hexachlorocyclohex-
anes and DDTs in pine needles. Environ. Sci. Technol., 37: 2350–2355.

Mattina, M.J.I., W. Iannucci-Berger and L. Dykas, 2000. Chlordane uptake and
its translocation in food crops. J. Agric. Food Chem., 48: 1909–1915.

Perez, T., E. Wafo, M. Fourt and J. Vacelt, 2003. Marine sponges as biomonitor
of polychlorobiphenyl contamination: concentration and fate of 24
congeners. Environ. Sci. Technol., 37: 2152–2158.

Prados-Rosales, R.C., J.L. Luque Garcia and M.D. Luque de Castro, 2003.
Rapid analytical method for the determination of pesticide residues in
sunflower seeds on focused microwave-assisted soxhlet extraction prior
to gas chromatography–tandem mass spectrometry. J. Chromatogr. A,
993: 121–129.

Simonich, S.L. and B.A. Hites, 1995. Organic pollutant accumulation in vegetation. Environ. Sci. Technol., 29: 2905–2914.

Takeuchi, I., S. Takahashi, S. Tanabe and N. Miyazaki, 2001. *Caprella* watch: a new approach for monitoring butyltin residues in the ocean. Mar. Environ. Res., 52: 97–113.

Thomas, G.O., J.L. Jones and K.C. Jones, 2002. Polychlorinated dibenzo-*p*-dioxins and furan (PCDD/F) uptake by pasture. Environ. Sci. Technol., 36: 2372–2378.

You, L., M. Sar, E. Bartolucci, S. Ploch and M. Whitt, 2001. Induction of hepatic aromatase by *p,p'*-DDE in adult male rats. Mol. Cell. Endocrinol., 178: 207–214.

Index